MACAO REMEMBERS

MACAO REMEMBERS

Jill McGivering

With photographs by David Hartung

OXFORD
UNIVERSITY PRESS

OXFORD
UNIVERSITY PRESS

Oxford University Press is a department of the University of Oxford.
It furthers the University's objective of excellence in research, scholarship,
and education by publishing worldwide in

Oxford New York

Athens Auckland Bangkok Bogotá Buenos Aires Calcutta
Cape Town Chennai Dar es Salaam Delhi Florence Hong Kong Istanbul
Karachi Kuala Lumpur Madrid Melbourne Mexico City Mumbai
Nairobi Paris São Paulo Singapore Taipei Tokyo Toronto Warsaw

with associated companies in Berlin Ibadan

Oxford is a registered trade mark of Oxford University Press

© Oxford University Press 1999

First published 1999
This impression (lowest digit)
1 3 5 7 9 10 8 6 4 2

British Library Cataloguing in Publication Data
available

Library of Congress Cataloging-in-Publication Data

McGivering, Jill, 1964–
Macao remembers / Jill McGivering; with photographs by David Hartung
p. cm.
ISBN 019-591735-9
1. Macao—History.   2. Macao—Social conditions
3. Macao—Relations—China.   4. China—Relations—Macao.   I. Title.
DS796.M2M38   1999
951.26—dc21   99-36869
CIP

Printed in Hong Kong
Published by Oxford University Press (China) Ltd
18th Floor, Warwick House East, Taikoo Place, 979 King' Road, Quarry Bay
Hong Kong

In memory of my father Ian Charles McGivering

All my love, pet

Acknowledgements

The greatest thanks must go to the people whose thoughts and memories form the substance of this book. This is quite simply a collection of their diverse stories, spanning rich and poor, political and social, some grand in scale and others very personal. My thanks to all of them for their kindness in agreeing to spend time sharing their stories and for their willingness to speak with openness and warmth to a stranger.

Many people have offered advice and suggestions, but a few deserve special mention. Heidi Ho of the Macau Government Information Services was an invaluable source of help, especially in the early stages of research. Paul Pun, the secretary general of Caritas in Macau, also offered great support with unfailing cheerfulness, despite the demands of his own considerable workload. Ana Brito at Macau's Maritime Museum provided generous help and information about the fishing community. Ana Borboa and Gloria Batalha Ung kindly gave freely of their time to help me with translations. Thank you to them all.

Contents

Introduction

Macao grows on you. It is often compared, not always usefully, with Hong Kong, usually to Macao's detriment and usually by people who live in Hong Kong. The two have obvious parallels. Both are southern tips of the China coast settled by past generations of ambitious European merchants, eager to trade with China and Japan. Both are places, part Chinese, part European, which soon developed their own colour and personality, influenced by the Chinese Mainland. Now, in the final years of the twentieth century, both are returning to it.

Given that their personalities are so different, it is hard to take the comparisons very much further. Macao is the grand older sibling, with four and a half centuries of Portuguese history behind it, compared to Hong Kong's one and a half centuries of British rule. Life in Hong Kong is fast paced, aggressive, overcrowded, and dynamic. Its global economic success, despite its lack of natural resources, is legendary.

Macao is very different. Its population is less than half a million compared to Hong Kong's six and a half million. Its personality is more thoughtful, amiable, gentle. Macao is a mass of contradictions, steeped in European heritage. The Mediterranean atmosphere, with its mosaic tiled squares, broad tree-lined avenues, and pastel-painted buildings set with verandas and wooden shutters, is far more obvious to visitors than the British influence is in Hong Kong. The Catholic Church, influential in Macao since the earliest days of European settlement, is still a key institution, making a major contribution to local education and social welfare. Macao is often called a City of God—as well as a City of Casinos.

Macao can also be read as a place very much in cultural transition. Its population is now dominated by relatively recent arrivals from mainland China. Nearly 40 per cent of the present population arrived after the mid-1980s, the vast majority from across the porous border with the Mainland. The population is generally more patriotic about China and less apprehensive about the return to the Mainland than the people of Hong Kong in the run-up to the territory's 1997 handover.

The contrasting pictures of Macao painted in this book hint at its present identity crisis. Those who cherish the Portuguese cultural influence want to preserve the elements of Sino-European cultural fusion as a permanent and essential feature of Macao; they see this as what distinguishes Macao from the rest of China, and what will guarantee its future. Others, already welcoming the return to China with open arms, see that legacy as history, part of Macao's past but not relevant to its future.

As well as its identity crisis, Macao also has an international image problem. For many people, Macao has become synonymous with the glitz and glamour of casinos and thrill-seeking of gambling. The casino industry is the central plank in its economy, a key revenue earner and tourist attraction. The much publicized wave of violent crime that has hit Macao in the last two years has superimposed an image of lawlessness and widespread gangster influence. A whole section of this book covers gambling and the criminal underworld from a variety of different perspectives. Elsewhere, other voices address the question of whether or not—and how—Macao should diversify its economy to make gambling less crucial.

The real story of Macao, however, is the story of its people. Macao is a rich and tightly knit community, which casual visitors find hard to penetrate. This sense of community has been the real discovery of doing the interviews for this book. Previously, as a weekend visitor to Macao from teeming Hong Kong, I used to enjoy the sense of tranquillity Macao offers but could not begin to understand the place. It was only as I travelled back and forth, week after week, to listen to these stories and reflect on the personal memories shared in this book that I started to gain insights.

I remember one visit, a warm day in late summer, when I raced across from Hong Kong on the jetfoil—the regular fast ferry service that connects Macao to Hong Kong in less than an hour's travel—after a hectic week's work. The contrast with Macao had never seemed more acute. There, people were strolling through the winding side streets, promenading European-style, not pushing and shoving as they had been in Hong Kong's busy business districts.

Small knots of people were lingering on street corners and in alleys, chatting and musing about the day. The quality of the colours struck me with force, as did the clarity of the light, illuminating the rich pastels of the buildings and white-washed brilliance of the churches with their small square bell towers. The market stalls, ponderous and friendly, were selling everything from oriental fruits and vegetables to bright red Chinese curios and good luck charms.

I rushed to the office of my interviewee, a well-known figure in Macao society, who politely made time for a foreign stranger to come and ask for his memories of the past. I sat entranced as he talked about the old Macao, the Macao of his youth, which has now disappeared forever, with its grand families, majestic sweep of the old harbour, and white rolling clouds. He spoke in a slow and gentle descriptive voice, looking into the middle distance as if he could see once again his family, his grandparents, the old Macao of the early years, reaching into the past with a lyricism and nostalgic yearning that I found deeply moving.

After the interview I had a few hours to wait until my next appointment and wandered into a small park, set about with shady trees, its focal point an endlessly gushing fountain. Clusters of elderly people, mostly men, had gathered along the shaded benches. The murmur of Cantonese conversation and strains of Chinese opera, played on a tiny cassette recorder, drifted through the air. Small children, toddlers, were tottering to and fro, minded by grandparents.

As the afternoon wore on, a slow procession of elderly men joined us, each arriving separately, each carrying one or more small bamboo bird cages. Each cage held a small, brightly coloured songbird. As the men arrived, they carefully took off the white cotton covers on the cages and hung them from loops of wire dangling from the trees, until the branches were as adorned as a Christmas tree covered with baubles.

The men sat, smoked, chatted; the birds twittered and sang in their cages; the toddlers played; the light fell about us all in dappled pools. I sat and tried to absorb it all, my head full of the thoughts and memories of the people who had spoken to me . . . their descriptions of the uncertainties and deprivations, but strange snatched joys, of wartime; of parents and childhood and near legendary ancestors; of the 1960s, with its riots and political turmoil, and the 1980s, with its sudden spurt of economic growth that seemed poised to transform Macao at last, after decades of post-war sleepiness. And I pondered their thoughts on the future—the uncertainties for Macao and the Macanese community under Chinese administration, and the sense of both endings and new beginnings that comes with the close of a chapter, a chapter which has spanned almost 450 years.

Political Portraits

Macao is a small place—with a huge political pedigree. It is the oldest surviving European territory in East Asia, dating back to the mid-sixteenth century. The return of the enclave to Chinese administration on 20 December 1999—fittingly enough, on the eve of the old millennium—brings to a close a centuries-long chapter of European involvement in the region.

It all started in the opening years of the sixteenth century. The Portuguese, with their considerable maritime power, had already established footholds in Goa and Malacca as part of a large and ambitious plan to dominate the spice trade routes. The traders were also eager to develop business with Japan and China, nations still considered in many ways mysterious by Europeans; such attitudes fuelled more by maritime stories and legends about the East than by documented experience.

As part of their strategy in the Far East, the Portuguese were keen to establish a permanent base in southern China, a convenient stopping-off point for travel north along the Chinese coast and also for through routes to Japan. The Macao peninsula was a prime candidate, a natural harbour, well situated to settlement and with sheltering islands. The Chinese response to Portuguese overtures about Macao was initially frosty but finally a compromise was reached. Sovereignty over the area was not conceded, but in 1557 China and Portugal signed an agreement allowing the Portuguese to settle there—a favour granted partly as thanks to the Portuguese for ridding the southern Chinese coast of a band of pirates.

So, unlike Hong Kong, Macao did not become European as the result of conflict with China. The Portuguese in Macao like to emphasize that difference in the origins of Macao and Hong Kong—one characterized by cooperation and the other by conflict. The cheekier amongst them will add that the two returns to China can be characterized the same way.

After that gentle start, Macao soon prospered, rapidly developing into a key staging post for Portuguese traders on their long voyages to the East and forming a lucrative link between China and Japan. Macao's political system was haphazard in the early days. Although important strategically, Macao must have seemed far too small and too far from Lisbon for the establishment of a formal system of government to be worthwhile.

The result was a pragmatic system of dual jurisdiction. The Portuguese governed their own affairs, using a Senado, or Senate, based on the model of local government prevalent in medieval Portugal. The Senado, later to become known as the Leal Senado, or Loyal Senate, comprised three councillors, elected by the Portuguese in Macao to serve for terms of three years. It had political and judicial powers—but only over Portuguese nationals. This early level of democracy led to Macao being described as 'the first democratic republic in the Orient'—a title in which Macao still takes great pride. The Chinese authorities, too, exercised jurisdiction in Macao alongside the Senado, intervening in disputes in the Chinese community and other areas of government involving members of the non-Portuguese population.

The rather muddled but pragmatic situation of parallel authorities started to be resolved at the end of the eighteenth century, when Portugal strengthened its hold on the area. By then the British had mounted a serious challenge to Portugal's control of the South China seaways. Portugal, struggling to keep its grasp on Macao during a series of attacks and disagreements, watched anxiously as the British founded Hong Kong just round the coastal corner in 1841, at the end of the first Opium War. Lisbon started to press for treaty terms with China over Macao as a way of securing its position. The two sides finally struck a deal and Macao formally became a Portuguese territory, an overseas province of Portugal.

Yet, after centuries of foreign rule, modern political events led to the eventual return of Macao to China, but raised serious questions regarding the autonomy, governance, and identity of Macao. A seminal political landmark of the twentieth century came in 1966 with the dramatic 1 2 3 incident, so called because the main events took place on 1, 2, and 3 December. Mainland China was in the grip of political fervour at the time, due to Mao Zedong's Cultural Revolution. Anti-foreign feeling was strong and the conditions were ripe for angry pro-communist uprisings in the foreign administered territories of Macao and Hong Kong.

In Macao, the spark was a local dispute about a pro-Communist school. The local authorities stepped in to stop construction work on the school site, on the grounds that the necessary permission had not been sought. The workers ignored the order and carried on building. Police intervened—but local people rallied to the workers' cause, pelting the police with bricks. The conflict escalated quickly. Riot police were deployed and soon vocal pro-Communists in the Chinese community were staging passionate rallies, waving their Little Red Books of Chairman Mao's thoughts, and denouncing the Portuguese authorities. Opinions differ about how spontaneous these displays of civil disorder in the local population were, and whether they were organized and encouraged through intervention from the Mainland.

The arrival of a newly appointed governor, Brigadier Nobre de Carvalho, who came new to Macao in the middle of the confusion, complicated the situation. The fervent anti-Portuguese demonstrations increased, and frequent violent clashes between police and demonstrators on the streets added to the sense of escalating chaos and instability. Some public buildings were damaged badly, and the number of injured began to rise steeply.

The events came to a dramatic climax when troops from the Portuguese garrison, called in to restore order but with little experience of handling civil conflict, opened fire on a crowd of demonstrators. The official account states that eight people were killed and 123 others were injured, although these figures have been disputed by the pro-Communist camp. One of the dead was a policeman, attacked by the crowd. A tense period of political stand-off followed. Many Chinese shops, stalls, and services imposed a boycott on the Portuguese. Leaders in the Chinese community, most notably a prominent Chinese businessman, Ho Yin, put pressure on the new Governor to agree to a series of demands, including an official apology and compensation to the families of the dead, which, by and large, he finally did.

Some commentators see the 1 2 3 incident as a crucial turning point in Sino-Portuguese relations in Macao—a forced subjugation of the Portuguese administration, from which it has never really recovered. Until that time, Macao had been viewed by some Chinese Communists as a dangerous haven for the Kuomintang (KMT, or Nationalist Party) forces. Many KMT supporters had fled to Macao in 1949, seeking shelter. Some on the Mainland accused Macao of harbouring Nationalist insurgents, saying the Nationalists were using the enclave as a base for fighting communism. The porous border between China and Macao, traditionally far easier to cross than the border with British Hong Kong, added to these suspicions. After the 1 2 3 incident, Nationalist activities in Macao were quashed.

Even today, there are conflicting versions of the 1 2 3 incident. Some key facts, such as how many people were hurt and how the shootings came about, are disputed, remembered differently by those in the Macanese and Chinese communities. More than thirty years later, it remains an emotional topic, a rare period of violence and instability in Macao's history, which caused enduring divisions in Macanese society.

Some local people say that it changed the balance of power between the Portuguese and the Chinese in Macao forever on a practical level, bolstering the status of the local Chinese and paving the way to fresh opportunities. Others point to the contrast between Macao and Hong Kong during the turbulence of the mid-1960s, suggesting the Portuguese administration's gentle, more conciliatory approach weakened its grip on Macao, while Britain's tough and uncompromising line on dissent allowed it to strengthen its position in its territory. The differences in perspective and in memory are best illustrated by the first-hand accounts that follow.

The second political turning point came in 1974 and was dictated by events far away in Lisbon. Portugal's revolution and the emergence of a left-wing government led to a complete change in thinking in Lisbon regarding foreign territories. The last outposts of the empire, including Angola and East Timor, were returned to local control. Portugal also tried to divest itself of Macao in the mid-1970s, but China refused to accept its return, saying the timing was not yet right.

Instead, a compromise was reached. The Portuguese and Chinese agreed that Macao was no longer under Portuguese sovereignty but a Chinese territory under Portuguese administration. A new era of local politics began, characterized by far greater autonomy and the beginning of democratization. Young, well educated Macanese, many then only in their thirties but spotted as potential future leaders, such as Dr Jorge Rangel and Mrs Anabela Ritchie, were enlisted to take part in the drafting of a new constitution for the enclave, the Organic Statute of Macau.

The new climate also led to the establishment of the Legislative Assembly, the legislative body in Macao, which would survive beyond the return to China. Its twenty-three members were to be chosen through three different methods. Eight members were elected directly. Eight represented functional constituencies, elected indirectly by closed groups representing business and other communities. A further seven were appointed by the governor. The governor himself, the head of the administration, was directly appointed by Lisbon.

The increased autonomy made it possible for Macao to make earlier, although only partial, progress towards local democracy than its dominant neighbour, Hong Kong. One challenge was persuading leaders in the Chinese community to take part in the new elections. Chinese representatives to the Legislative Assembly for some time were more visible in the functional constituency seats than in the directly elected ones, despite Chinese predominance in the population.

Some people suggest this slow, but incremental, development of local democracy since the mid-1970s made it a less acute political issue in the run-up to the return to China. The apparent lack of fervour about democracy certainly makes an interesting comparison with Hong Kong, where democracy became a passionate rallying point in the approach to Hong Kong's return to China in 1997, and where support for the Hong Kong Democratic Party remained strong following the handover. One explanation is that Macao lacks a large, well-educated middle class, the obvious champions of democratic reform. Only recently has Macao offered a range of opportunities in tertiary education and the sort of white-collar jobs that would entice those who could afford an overseas education to return to Macao.

The sense of uncertainty and apprehension felt by many people in Hong Kong in the run-up to 1997 has been largely absent in Macao's local Chinese community, which forms the overwhelming majority of the population. The local Macao Chinese are often described as more patriotic about China than Hong Kong people. This sentiment might be partly because a large percentage of local Chinese in Macao only settled there some ten to fifteen years before the handover, with the opening up of China in the 1980s. As a result, many see themselves as Chinese people who happen to be living in Macao; the return to Chinese administration is often described as a natural evolution rather than an alarming threat.

The next key date in Macao's history is well known: 20 December 1999—the return of Macao to China, under the terms of the Sino-Portuguese Joint Declaration signed in 1987. The model for its return mirrored that of Hong Kong, whose own Joint Declaration was signed in 1984. The declaration stipulated that, like Hong Kong, Macao would be a Special Administrative Region of China, with autonomy on local affairs but under the auspices of the central government in Beijing, which would retain control over national defence and foreign affairs.

Macao's post-handover constitution, the Basic Law, guarantees that its way of life and pre-handover legal system will be preserved for fifty years. The Legislative Assembly remains largely intact through the transition, notwithstanding a few changes to key positions such as the Assembly's president. The position of governor of the Portuguese administration has been replaced by a chief executive. The first Chief Executive, Edmond Ho Hau-wah, a prominent local banker and the vice-president of the pre-handover Legislative Assembly, was chosen by a special committee whose members were drawn from Beijing and the local Chinese community in Macao.

Dr Jorge Rangel
Macanese Leadership

Macao's mixed-race community, usually referred to simply as the Macanese, is a unique feature of the enclave's development. Some European powers frowned on social integration when they built colonies, and children who were the products of marriages between Europeans and non-European local people often found themselves outcasts, not fully accepted by either community.

Portugal's record is quite different. Intermarriage was often seen positively as an important step towards social harmony. From the days of the first settlers in Macao, Portuguese sailors and merchants married women from various reaches of the Portuguese empire, from Angola and Goa to Japan and Timor. Intermarriage with Chinese was less common in the early centuries, when the social barriers were more rigid, but became more so in the last hundred years.

The distinctive cultural legacy of the Macanese is considered later in this book, but their political contribution has been significant, too. They provided the answer to an awkward practical problem. Portuguese was seldom taught in Chinese schools. Portuguese officials seldom learned to speak the local Cantonese language or read and write Chinese characters. This obvious gap in communication was often bridged by the those within the Macanese community who straddled the two cultures and were proficient in both languages.

As a result, the Macanese community had considerable influence and status, often identifying with the Portuguese way of life but drawing on a Chinese heritage as well. Macanese also played a key role in Macao's civil service, providing language skills and an important element of continuity for the Portuguese administrators.

Historically, the Macanese have a tradition of emigration. Until recently, many Macanese had to study overseas if they wanted the tertiary education that Macao could not then provide. In the days when Macao was still a sleepy enclave, dominated by its fishing industry and a small manufacturing sector, many young Macanese saw the career opportunities in Macao as too limited and chose not to return when they graduated. As a result, the Macanese diaspora is large and widespread, with associations as far afield as Brazil, the United States, Australia, and Canada.

Dr Rangel comes from a well-known Macanese family that has been in Macao and other parts of East Asia for ten generations. His influential role in the last Portuguese administration, serving as Acting Governor many times in the Governor's absence, stemmed from his own position as Secretary for public administration, education, and youth, a role that included overseeing the localization of the civil service in the run-up to the handover and the rapid development in recent years of new educational opportunities in Macao, especially at the tertiary level. Dr Rangel speaks with passion about the development of education in Macao, something he sees as crucial in building the enclave's future.

Dr Rangel began his political life at a time of great change, just as the Portuguese revolution was paving the way towards greater political autonomy in Macao. In 1976, he won a directly elected seat in the first Legislative Assembly, set up under the new political system. He is a high-profile example of the young generation of Macanese, now reaching middle age, who returned from overseas as recent graduates in the aftermath of the revolution to take part in the task of building a new political structure for a new Macao.

WE FEEL WE BELONG HERE and when we talk to the Chinese authorities, we remind them we have the right to say something about the future of Macao. We've earned the right to take part in the changes taking place now and to play a role in the decisions being made, whatever the future may be.

The first ancestor of mine to come to Macao arrived in the seventeenth century. He was a member of the original Macao Senate and in charge of the Holy House of Mercy, a very important institution historically in Portugal, which was designed to help the poor and destitute. He was a merchant and when he died he left all his wealth to the church. That happened very often. I remember two aunts of mine, neither of them had married and they, too, left everything to the church. So it was still happening, even in my lifetime.

One Portuguese historian described Macao as the first democratic republic of the Orient, a very interesting title. But only the first few of my ancestors were involved in public affairs. The others were mostly merchants or businessmen. On my mother's side, my grandfather came from Portugal as an army officer and was the chief of police here for many years.

I had a most interesting childhood. I was one of the fortunate few who had everything I needed. I used to live in an area where many of the Macanese grouped together, near the hockey field. The mansion we lived in has now been converted to house the Macao archives collection. The office now used by the director of the archives—that was my bedroom. My childhood days were spent almost entirely in that small part of town. My primary school was just across from our house, across the field, and my pre-school as well. My secondary school was also close by, so my whole world as a child was right there.

Macao was very quiet in those days—life was completely different. There were few cars. We used to play football on the main road, something that would be impossible today. We'd see a car coming from a long way away and stop playing for a few minutes to let it pass. We knew all the owners of the cars—we'd say: 'Oh, look, Mr So and So's coming.' We spent a lot of time on the streets and it was a very happy childhood.

I used to collect toy soldiers, made out of lead. I had a huge collection. Ours was a vast house and on Saturday or Sunday I'd parade all my soldiers right along the corridor from one end of the house to the other and spend hours playing with them all. Or my friends would call for me and we'd go and play field hockey and other games. I used to play goalkeeper, the same position my father played when he was a young man. I had a bicycle and would go riding in town.

Our household was made up of my grandfather, until he died, my grandmother, my father, my older sister, and me—just the five of us in that huge house. My mother died when I was four. She was in labour with one of my sisters who died at the same time. I only vaguely remember my mother. I remember her face—but that may be because I've seen many pictures of her since then. My birthday fell a few days before she died and she gave me a wooden rocking horse. I remember riding on the horse when she was already in hospital. I went there every day to visit her and we left the rocking horse at the hospital, in her room, so I could play with it when I was with her. Then I was suddenly told I couldn't go there anymore. I didn't know she'd died. It was some time later that I realized she'd gone for good.

My father was still young when he lost her. He worked for a private company here, representing British Airways and some shipping companies. His job meant a lot of travel so I was often left with my sister who was a year and a half older than me. We had a devoted servant from China who took care of us. She was like a second mother and stayed with our family until she died. Those were the days when you could find people who were really dedicated—to their work, to associations, to their country, to family. That's how Macao used to be.

It was typical for the old Portuguese families to live in good houses, particularly the mansions along the bay. We moved quite a few times, once to Central Street, one of the oldest commercial streets in Macao where we had a big house with four doors. I was ten or eleven, immediately after my father married again. A few years later, we moved again, to a house close to the Government Palace. Every day when I left the house, I'd see the bay and always strain to look beyond the horizon. It stirred some urge in me to go out there and see what was beyond that horizon. Even now, I can only feel happy if I live near water.

I left Macao in 1963 to study in Portugal. It was my first trip to Europe and I had no idea where I was going. In those days it was very difficult for people to study beyond secondary school. There was no university, no institution of higher learning of any kind, so if you were ambitious and had the means, you had to go elsewhere. In my school days you either had to be rich or have a government scholarship to study. I received a government scholarship. My family could have supported me but it was the scholarship that made me decide I had to go away and study. It wasn't a loan but an award and my father told me: 'You have to go.'

I'd been pressing my father since I was fifteen to leave Macao and see the world. Because he was working for a company that represented shipping, I told him I wanted to work on a big ship and travel across the globe. He said: 'You've got to get yourself ready for the job, you can't just join a ship without any qualifications.' So I studied for a diploma as a telegraph operator while I was still finishing secondary school, a two-year programme that I finished when I was seventeen.

I went back to my father and said: 'Right—it's time for me to go now, I've got my diploma', and he said: 'No, you've got to finish secondary school.' Then, finally, I won this scholarship and everything in my life changed. I also owe a lot to my father's, and to my eldest sister's, encouragement and support.

In those days, people who left Macao never knew when they'd come back. I was away for twelve years. A friend of mine was away for twenty. Portugal was very different then. It had many overseas territories and I always felt I was the product of the country called Portugal, which stretched all the way into Asia. For me, it wasn't important to come back to Macao. I just wanted to be in some part of my country, and that meant Portugal.

So when I was called to join the army for compulsory service, I thought I would join for a couple of years, the normal period for army officers. Then suddenly I was put in command of an infantry company and sent to organize a company in the Azores. There I was, someone from Macao, studying in Portugal, going to the Azores to organize an infantry to fight a war in West Africa. That was the concept of the country. It wasn't so much a European country as an African one in those days.

I was prepared to stay on in Africa, not in Guinea Bissau but Angola, which was the real challenge for my generation. When we had the revolution in Portugal, I was still in Guinea Bissau. I started off in command of this infantry company, then the General called me to work with him in his office there. I used to travel often by helicopter, visiting various military garrisons—an excellent experience for me. It was an independent company, which meant it wasn't under the battalion but completely separate. The boys were all very young, between sixteen and eighteen. I had to be everything to them—mother, father, priest, doctor—the lot. They were all volunteers who wanted to emigrate to the United States or Canada and they joined the army early in the hope of emigrating as soon as they left.

I was called to the army when I was twenty-seven. If you were studying, you were allowed to defer your military service, so I studied first in Portugal, then Cambridge, then Bonn, until I couldn't defer any more. I studied languages and literature—basically Germanic studies, which was why I took my post-graduate degree in Germany. I also had a German grandmother, on my father's side.

I just caught the aftermath of the 1 2 3 incident in 1966. I came back for the Christmas holidays and by coincidence flew into Hong Kong on 4 December. I had no idea what had been going on. My father came to Hong Kong to meet me at the airport and said: 'We can't go back to Macao. There's a curfew.' I was astonished because Macao was always so quiet. But my father explained there were riots and told me we had to spend the night in Hong Kong and go back the following day.

I found everything quite changed here. You could feel the tension. My father's office was on one of the main streets and just next to it there used to be an Australian restaurant, called 'The Waltzing Matilda'. It was closed because of the disturbances but there were chairs left outside and I used to sit there every day with a book, waiting for my father to leave the office so we could walk back home together.

There were very few people out and about at all. Some people had been killed when the army was called in, because the police were unable to contain the situation. When you call in the army, anything can happen. One of the soldiers, just one, opened fire on the crowd of demonstrators who'd pulled down one of the statues and were destroying buildings round the square.

When they held the funeral for those who died, I was the only person out on the street when the procession went by. I was sitting quietly outside the restaurant with my book. People warned me and told me to go home, but I decided to stay and see what happened. In fact, no one bothered me or challenged me. I just stood there watching the funeral procession pass by. Soon after that, the government signed an agreement with the Chinese to bring things back to normal. But the riots did shake Macao, and many people left. They were really a result of what was happening in mainland China, with the Cultural Revolution. It only needed one small spark here for the chaos to spill over into riots in Macao.

That spark was a new school in Taipa Island. Construction wasn't allowed to start until the pro-Communist builders received a licence from the government. But the construction of this school went ahead, in defiance of the authorities. When the police stopped them, riots broke out. It was a time of complete craziness in China.

I felt no empathy with the people rioting. They were Red Guards, wearing red armbands, parading through the streets of Macao, shouting slogans from Chairman Mao's Little Red Book. They used to congregate in the workers' sports ground near the Lisboa Hotel and march from there. I thought it might be the end of Macao. I think that's what most people thought.

When I returned in the 1970s, Macao was still quite quiet. There were more people, more cars, more businesses opening, more life in Macao; but there were also questions about the future. Portugal gave independence to its overseas territories in 1974 and 1975 so now the big question was: what about Macao? I hadn't intended to come back at all. I came to visit my father for a holiday, and the Governor called me to his office and asked me: 'Why won't you come back and work here? You were born here.' It was a challenge. In the end I told him I would come back, but only for two years. In fact, it's been almost twenty-four years now.

At this time, the end of 1975, we were just completing the Macao Organic Law, which established the principle of elections for Macao's first Assembly. In December 1976, the Organic Law was published and elections were announced for June that year. A group of young Macanese people decided to get together and offer the voters another choice, just to see what might happen, and I became one of the candidates. Surprisingly, we came in second in the elections. We were young and little-known in Macao, but we offered a new message and new ideas. So I found myself an elected legislator.

I was the director of tourism and government information at the time and went to talk to the Governor. I suggested these two roles were incompatible. He was a young, open-minded man who had become Governor when he was only thirty-four. He said: 'I'm sure you can tell when you're performing in one role and when in the other, so why don't you carry on and see how it goes?'

I told him I'd have to have the right to voice my opinions freely in the Assembly because of these responsibilities, and he agreed. It was a very interesting time of my life, being an elected member of the Assembly and a senior government officer at the same time. One key task was to encourage more people to register to vote. In the beginning, most of the people who voted were Portuguese. The Chinese weren't really interested. To start with, we had less than four thousand registered voters. Now we have more than one hundred thousand. Of course it's still not enough but it's a big difference.

It was an exciting time. We felt we could really change things. Until the Portuguese revolution, all the major decisions were made in Lisbon by the Overseas Affairs Ministry. If you wanted to build a bridge or a school, it had to be approved by Lisbon. But after the revolution, Macao was given a high degree of autonomy, and the governor was allowed to make these sorts of decisions himself. That's why we created the Assembly, so we could pass the laws ourselves. That process has lasted all the way to the handover itself.

When another governor was appointed in 1981, he asked me to join him in the government and I became Secretary for tourism, culture and education, and left the Assembly after completing my four-year term. I stayed in that job until 1986 when there was an election in Portugal and the new president chose a new governor. It was a very strange decision. He'd never been to Macao before, not even as a tourist, and knew nothing about the place! He was a professor of medicine who came to Macao just for one year. Although a distinguished gentleman, he had no past connections with Macao and brought along senior officials from Portugal who also came for the first time. I strongly criticised the central government's political decision, which had nothing to do with Macao.

That was when I got married. I called my wife and said: 'Let's get married quickly because I don't know the new governor and he doesn't know me—this is my chance to have some time away from the government.' It was an excellent year; my only daughter was born and we all had holidays together, weekends in Hong Kong—it was a wonderful time.

Then there was another new governor, someone I thought should have been appointed a decade earlier. We had been suggesting his name to Lisbon since 1981, but he only came in 1991! It was the right choice, but made too late. We raced to try to accomplish in less than ten years everything that had been neglected for decades. For example, the education system should have been developed much earlier. One of our main problems now is that although we have very good people, they're young—in their late thirties or early forties—and inexperienced.

There are advantages to youth, especially at a time of change, but even so. . . .

I was invited to become the first president of the Macao Foundation, which was founded to develop higher education in Macao. I was given responsibilty for changing Macao University, then a small private university, into a major public institution. The university was an essential tool in training the young people we needed for Macao. Once that was done, we created the Macao Polytechnic, of which I was the first president, and persuaded private enterprises to start their own Open University to offer distance education for those who had no chance to study when they were younger.

We also opened an Institute for European Studies, which has strengthened Macao's European identity, and another academy to train local senior police officers so they can gradually replace the police officers who came from Portugal. The United Nations University also opened one of its specialized centres in Macao, and the Portuguese Catholic University established an Institute for Advanced Studies in a joint venture with the Macao Diocese. Now we have more than two thousand Macao students studying abroad, as well as thousands studying here in Macao. I'd say this has been our major investment in the last ten years—but oh, we should have started much earlier!

The Macanese community regard themselves as Portuguese and some Chinese people do as well. I was just reading in the newspaper about a man who is ethnically Chinese but culturally Portuguese. He says his heart is in Portugal. He wants to carry on being Portuguese and he's very worried about the nationality issue.

Of course, on Portugal's side, there's no problem with that. Everyone with a Portuguese passport is a Portuguese national and entitled to all the rights offered to any Portuguese citizen. So in Macao there are very many people, about a hundred thousand of them, including ethnic Chinese, who have no direct connection with Portugal but carry Portuguese passports. It's exactly the same passport as the one issued in Lisbon, which means these people can settle freely anywhere in Europe, as European citizens.

But on the Chinese side, there is a problem. China doesn't accept dual nationality. The Portuguese say you can be Portuguese and Chinese at the same time. But the Chinese nationality law says you can only be a foreign national if you go and live abroad, and acquire a foreign nationality—then you cease to be Chinese. China will allow people in Macao to use the Portuguese passports as a travel document, but as far as nationality is concerned, people are Chinese.

The problem here is the Macanase. How does one decide who is ethnically Chinese and who isn't? We argue that the Macanese should be allowed to choose to be Portuguese even if they have Chinese blood. It sounds easy to resolve but we're really running into problems. The Chinese say all those who have Portuguese blood may be regarded as Portuguese or they may choose to become Chinese nationals. But what does one really mean by Chinese or Portuguese? What about the man I mentioned who is ethnically Chinese but considers himself culturally Portuguese? We want everyone in Macao to be allowed to choose between one nationality or the other.

As for me, I feel I belong here, but culturally I am, of course, Portuguese; we've been Portuguese for generations. My values, my way of life, the way I feel, the way I look at things—they're all a product of my European education. The Basic Law says very clearly that the residents who are descendants of the Portuguese will be protected by law. Something must be done to protect the interests of this small community that is a fundamental part of Macao and very much responsible for shaping its history.

Our future may depend on how many Macanese stay here, and how far they'll be united in defending our interests. We already have our own associations here and we've decided to establish a school that will continue to use Portuguese as a medium of instruction, supported by the ministry of education. There will be other clubs and associations that will help the Macanese community to have a voice. We hope the new government will understand that it is important for Macao to keep this community here. It's an issue of continuity. The Chinese in Macao are also different from the Chinese on the other side of the border. People here enjoy full rights, freedoms, and liberties. We feel the best thing Portugal can leave behind is our values—the way we think about law and order, respect for the individual, the rule of law.

As for fears of an exodus of Macanese after the handover, the Macanese have always left—that's not a new thing. Many Macanese emigrated to Brazil, Canada, Australia, the United States, and of course

Portugal itself—that has happened for generations—but just as many have stayed. I agree things may be more difficult in the future because there will no longer be a Portuguese administration.

Also, the influx of people coming to Macao from mainland China has almost doubled the population in the last fifteen years. These Chinese people—from various parts of China, looking for a job, for a better way of life, for economic opportunities here—have no roots here and that makes it hard to keep the identity of Macao. So we rely very much on the young generation of Macanese, and young Chinese who feel they belong here and understand this is a time of change.

I think we have to revise the whole concept of Macanese. Not only those who see themselves as having a Portuguese heritage are Macanese, but the Chinese people who belong here are, too. If we don't keep that sense of belonging, that something special, which is peculiar to Macao, there's no reason to maintain one country, two systems; this might as well all be China. If we want Macao to be a separate entity with its own values and traditions, we need younger generations to think in terms of being a new Macao community. Fortunately younger generations have this sense of belonging because they are also much more qualified, academically and culturally.

I've promised my family I'll have more time to be with them soon. In the last few years, I was given more and more things to do, on top of my own huge portfolio. Far too many things! I hardly see my family. I feel my mission will be accomplished on 19 December 1999. I will be available to give assistance to the new Chief Executive on a personal basis but I'd prefer not to be in an executive position in government. I'd rather return to academic life. I always believe no one person should stay too long in the same job, especially in a senior position. There are new people who will come in with new ideas, new plans. It will give me great satisfaction to pass on my responsibilities to a local person who is identified with Macao's past and present, and who has a broad vision of the future.

Mrs Anabela Ritchie

Pre-handover President

Anabela Ritchie, the pre-handover President of Macao's Legislative Assembly, is a high-profile example of the young generation of Macanese who played an important role in defining Macao's modern political structure. These young people returned to the enclave as recent graduates in the mid-1970s when Macao's political system was being redefined in the wake of the Portuguese revolution. The revolution sparked new ideas in Lisbon about its relationship with overseas territories—and brought new opportunities for more governmental autonomy in Macao.

Born in Macao, Mrs Ritchie comes from a Macanese family with both Portuguese and Chinese ancestry. Her parents, also Macanese, were both civil servants. Her grandmothers on both sides were fully Chinese, one from Guangzhou and one from a traditional Catholic Chinese family that had lived in Macao for several generations. One grandfather was a Portuguese military man who came to Macao at the turn of the century, settled, married, and died in Macao. The other came from mixed ancestry, with a strong Asian heritage.

Like many Macanese, Mrs Ritchie was brought up with Portuguese as the dominant culture and language in the home. She studied in Portugal as a young woman and began her professional life as a teacher there. After about a decade away, she came back and soon found herself beginning a hectic political life in the place of her birth. She was an appointed member in the historic first chamber of the new Legislative Assembly, set up in 1976 as part of the process of devolving power to Macao from Lisbon.

WE GREW UP SPEAKING mainly Portuguese at home. Some time after I was born, my parents decided my mother should stop working and stay at home to educate us. At that time there were four of us, and they felt they shouldn't leave us with servants all day. My mother had studied far more than girls usually could in those days and really enjoyed working. I can imagine how sorry she felt when my parents decided she should stop. We all feel we owe her a lot for doing that. She taught me to read and write. I didn't go to formal school until I was seven, which was unusual.

Her father was Portuguese and her mother Chinese but I'd say she considered herself much more Portuguese than Chinese. She always felt it was very important for us to have a strong sense of identity and at that time people were much more oriented towards Portuguese than Chinese culture. Our schooling was all in Portuguese, we spoke Portuguese at home, and although we spoke some Chinese, I'd say mine was very rudimentary. I'll never be able to say I'm really proficient in Chinese—my first language will always be Portuguese, my second is English, and Chinese is only my third. It's a real pity because Chinese culture is an important part of my identity.

Chinese was taught at school but it wasn't compulsory. I started learning French when I was ten, English when I was twelve, and only started learning Chinese in my twenties when I was already a mother. That wasn't a well balanced education. The contact we had with the Chinese community wasn't great. I had Chinese friends and many Chinese relatives, and spoke Chinese with them, but the big influence was certainly Portuguese.

It was different for my children. By the time they were growing up, Macao was much more Chinese than it was in my childhood, and they started learning to read and write Chinese when they were little. But today they're much more fluent in Portuguese, English, and French than in Chinese, so the pattern is being repeated.

On my father's side we've lived for five generations in a predominantly Portuguese area, around St Lawrence Church. Our education focused on Portugal. Thinking about it now, I might call it foolish, but when I went to Portugal at the age of seventeen, it was a world I already knew and recognized. The first time I travelled by train, I knew exactly when and where to get off and change trains, even though it was somewhere I'd never been before. There's such a gap in our knowledge of Chinese culture. Everything we acquired, we learned in later life.

There's been much debate recently about the issue of identity. A young friend said to me: 'How can I be Chinese? I might have a Chinese background but they're telling me to sing the Chinese National Anthem and I can't—I don't know it!' I said: 'I can't even pray in Chinese! Some years ago I went to a Chinese church and I knew people were praying and I could follow it, but I couldn't say the exact words.'

I'd been in Portugal for six months when the so-called 1 2 3 incident happened. Of course it was a worrying time. Everything was so far away, we didn't really know what was going on. We didn't know if things would change in Macao. My friends and relatives back in Macao were worried, too. I came back to Macao on holiday in 1968 and there were still signs of anxiety, but from then on, things started to settle down again. Everyone was very careful; they didn't know how the situation in China was going to evolve. I knew radical changes were taking place in China but didn't know what they would lead to. All we could do was wait and see.

I'd planned to stay in Portugal. Like most young graduates, once we started living in Europe, we became fond of everything we could do there, things we couldn't do in Macao. At the weekend, we could go out, leave town. I like big open spaces, getting in the car, and driving somewhere faraway. You can't do that in Macao. And the anonymity . . . in a big city, you mind your own business. You can go where you like and meet all kinds of people. For someone who lived all her life in a place as tiny as Macao, I really appreciated all the things a big community can offer.

I started dating my future husband in 1968 and we got married three years later. He's Macanese, too, and I'd known him all my life; we went to the same school and were in the same class. Portugal was fighting a colonial war at that time and when you graduated, you had to serve compulsory military service. My husband graduated as a medical doctor and, for a while, we were supposed to go to Mozambique, but then the war ended. The authorities approached him and said Macao needed a military doctor, and he jumped at the chance.

After returning to Macao, we began to be aware of the need for young graduates like us. We were only the second batch of Macanese graduates who were lucky enough to study in Portugal and get a tertiary education. The Governor of the time, Garcia Leandro, was keen to bring back young Macanese graduates and give them posts of responsibility. He used to travel and seek out graduates and personally ask them to come back. It was a wise policy, and that's why we decided to stay on in Macao. A new awareness had been born here. Macao was so in need of human resources and technical help.

I taught in a secondary school at first, teaching Portuguese, English, German, history, and a little bit of literature. It was still a European-style syllabus and, incredible as it sounds, some of the books I was given to teach were the same ones I'd studied as a girl. Some of the content was antiquated, but I got the chance to meet young Macanese who were very different from us. They had a civic awareness; not always very strong, but some had an acute sense of society and politics.

We came back right after the 'April 25th' movement, which marked the beginning of democratization in Portugal. It was the opening of a new chapter in Macao as well. The Legislative Assembly that existed until 1974 was presided over by a governor. There was no principle of the separation of power. Executive and Assembly worked in the same building and on certain days of the week, the governor would come down and preside over the Legislative Assembly.

That Assembly was dissolved right after the *coup d'etat* in Portugal, and work began on drawing up a new constitution. I was asked to join the council that was formed to assist the governor at both executive and legislative levels, in the absence of a legislative organ. Our work led to the approval by Lisbon in 1976 of Macao's Organic Statute, which is our constitution. For the first time, we saw the separation of the Legislative powers into an Executive and a Legislative Assembly, two-thirds elected. The Assembly's president, vice-president, and first and second secretaries were to be elected internally.

It was a wonderful, exciting time. I was very conscious that we were making history. My husband and I were both part of the group GEDEC [Study Group for the Community Development], which took part in the elections for the first Legislative Assembly under the new terms in 1976. I wasn't a candidate but we managed to get one person elected from our list of candidates, all of them young Macanese professionals. The governor had the right to appoint seven members of the Assembly and I was invited to join, to carry on with the work I'd started in the council.

It was a huge first step. For the first time, Macao had autonomy—administrative, economic, legislative, and financial autonomy. We were aware we had to work hard because we were no longer governed by

Lisbon, in the way we had been in the past. It was a new chapter in the history of Macao. Even in this new Assembly, there were more Portuguese than Chinese. At that time, the Chinese weren't very interested in elections. They saw elections and the electoral system as Western ideas, something for the Portuguese. They preferred to enter the Assembly by appointment, or through the functional constituencies. That changed. As time passed, they became more and more interested in being directly elected.

I was only twenty-six when the Assembly was formed. I felt young, but the Governor himself was only in his thirties. There was a real willingness to introduce new blood. That Assembly had four women, but I never felt being female was an issue. We shared a certain camaraderie in the Assembly, regardless of whether one was male or female.

I was aware I just couldn't give as much time to my children as my mother gave me. All my life, I have tried to find a way of reconciling being a mother and being a professional. My younger son was so used to seeing me busy that when he was little, he told people that when he was born, everyone was there, his grandparents, cousins, his father, and so on—but not his Mum because she was in the Legislative Assembly. He was convinced it was true.

My main goal was to do something in education. Teachers were paid less than other professionals. We felt this was unfair because we were graduates like everyone else. Legislators did manage to change that remuneration system. We also drafted legislation to subsidise non-lucrative schools, starting what led to the network of free schools that exists today. Legislation also stopped the selling of pornography to young people. Much of our focus was on welfare, education, and culture.

In 1992, the President of the Legislative Assembly passed away very suddenly and we had to elect a new one within fifteen days. People started discussing who could fill his shoes—in that debate, I never heard anyone talk about the fact that I was a woman; it simply wasn't an issue. I think that's a sign of intellectual and emotional maturity. I took up the post at that time, and I have been re-elected twice since then. I wasn't directly involved in drafting the Basic Law in the 1980s, but we all had some input. We focused on doing what we could to preserve the traditions of the Portuguese community. It's a small group of people but a significant one and, knowing 1999 would mean changes, we wanted to keep the characteristics of the community as a whole.

Many Macanese are leaving Macao. Macao is a small place and it's hard to develop a career, so the Macanese have always emigrated to greener pastures. All Macanese families have someone who has left. I have relatives in Hong Kong, Brazil, Canada, Portugal. We're used to it. So I'd say 1999 will increase that emigration, because some people will feel their children will have better opportunities elsewhere. My brother and sister left years ago and live in Portugal now. So do my children. Many Macanese families are repeating this pattern. Chinese characteristics will tend to dominate Macao in the future, but the Portuguese have been here for four hundred and fifty years—they won't disappear overnight. Some will stay on because they feel their roots are in Macao. My feelings are rather mixed but I shall always feel I belong to Macao. History is something you can't change. My own plan is to stay, and there are others like me. There'll be fewer of us, but we'll certainly carry on as a community.

The Sixth Legislature will end by 19 December 1999 and the First Legislature of the Special Administrative Region will begin then. The Basic Law is very clear, in that some key posts of the SAR have to be filled by Chinese nationals. The Basic Law is a Chinese law; it was felt that that's how it should be, and I can understand that. On the nationality issue, the Standing Committee of the National People's Congress of the People's Republic of China approved a resolution concerning the application of the Chinese Nationality Law in Macao when Macao reverts to China. According to that ruling, people of mixed ancestry (Portuguese and Chinese) are entitled to opt for either of the nationalities—the one they already possess (Portuguese) or the one they have according to the Chinese Nationality Law (Chinese nationality).

There will be other political changes, too. At the moment we share the power to legislate with the governor. That will cease after 1999, and the Legislative Assembly will be the sole legislator. At the moment, we don't formally approve the budget but after 1999, we will. There will also be changes to introducing private bills. I'd say our powers are greater now than they will be after the handover. But sometimes everything depends on how you use the powers you have. For more than twenty years, the Legislative Assembly has been a place where we could discuss matters openly. People from all walks of life have brought their problems and issues here, and it's been a very open place for dialogue. I'd love the Assembly to carry on in that spirit. The key is how willing people are to exercise their power within the Legislative Assembly.

Macao is certainly becoming ready for more democracy. In 1976, people were only just starting to develop civic and political awareness. It's been a long process. People are becoming increasingly interested in taking part in elections. Democracy can't happen overnight. I don't favour abrupt change. Many of us did not expect Chinese army troops to be here in Macao after the handover, but the important thing to me is that the Basic Law is enforced. That means that China is responsible for defence and the Special Administrative Region government is responsible for internal security. As long as that's strictly obeyed, we'll be all right.

When I was a little girl, I never dreamed I'd have such an interesting life. I wanted to study law but there weren't female lawyers at the time, and my parents persuaded me not to read law. They said it was a man's world. They said I'd have to go to court and deal with divorce—we were a very Catholic family—and take care of criminals. So in the end I read literature and languages. But it's an interesting twist because here I am, finally part of the world of law after all!

What will I feel at the time of the handover? I think I'll feel a chapter of history is closing. But Macao will always be my homeland. It's not the end, it's the beginning of something different. Something new. A new chapter of history will unfold, and my wish is certainly that it will be one I like.

Dr Sales Marques

Mayor of Macao

Dr Sales Marques is the de facto Mayor of Macao, the chairman of Macao's Municipal Council. While the Legislative Assembly approves legislation and enacts policy, the Municipal Council runs the day-to-day operations of the main area of urban Macao, from collecting fees and rates and controlling traffic to organizing the collection and disposal of rubbish.

He commands the majestic surroundings of the Leal Senado, one of Macao's grandest old buildings and an architectural landmark in the centre of Macao. It stands at the head of Leal Senado Square, a well-known reference point for tourists and locals, complete with Mediterranean-style mosaic pavements, fountain, and neo-classical façades.

The Leal Senado, literally the Loyal Senate, has an impressive history. It was first formed in the late sixteenth century, in response to the need for some early system of government in Macao and, until the modern system of government was established this century, it had great authority. It supposedly won its title in the seventeenth century when Spain occupied Portugal. The Senate in Macao refused to recognize Spanish sovereignty or to fly the Spanish flag during those troubled years. The original building, thought to have been built on the same site, is said to date back to the 1580s. The basis for the present building was established in 1783, although there was major reconstruction work in the 1870s after the building was badly damaged by a typhoon.

Dr Sales Marques comes from a family of Macanese civil servants who settled in the enclave several generations ago. His father worked in the post office, his mother was a general administrator with the government. Dr Sales Marques was born in 1955, when Macao was still recovering from the impact of World War II and struggling to rebuild. His parents were living with his grandmother at the time of his birth but soon afterwards moved to a house along the famous Praia Grande. It is this place he fondly remembers as his childhood home.

I HAVE VERY HAPPY MEMORIES of the old Praia Grande bay. Its beauty was legendary. The fishermen's junks could come in very close to shore and there were always lots of them. The sight of them on the water with their sails unfurled was a typical image of Macao in those days.

I grew up in a quiet neighbourhood, with few cars on the roads, so we could play in the street. We had our own house, quite a big but old property. We had trees in our back garden and plenty of space to move around. It was typical of Macao. If you look at the pictures of the Praia Grande from the 1950s and early 1960s, you see many two-storey houses along the bay. I lived in one of those. We had one live-in servant who stayed with us for more than sixty years. Sometimes we had a second servant for a while, various people who stayed for a few years at a time and then moved on.

I lived there until I was twelve or thirteen. I didn't go to kindergarten. We shared a private tutor with a few other families so I learned to read and write with him. At school we were taught the traditional Portuguese syllabus. We learned all about the other Portuguese colonies—I know the names of all the rivers in Portugal and Angola, but very little about Macao.

We had some tough times in the 1960s. The 1 2 3 incident made a strong impression on me. We lived just a stone's throw from the Government Palace. On the day it all started, I was at school near our home. My mother used to work in the same building, in rooms used by the education department. One of our friends called into my mother's office at about noon and told her to pick me up. The friend drove us both home and told my mother not to leave the house. That afternoon there was a riot near the Government Palace and the rioters fired smoke bombs. We were so near Government Palace, we could smell it in the air. We had to close the windows to keep the smoke out. I was young—about twelve years old—and I'd never seen such violent scenes before. It really shocked me. My father was told to stay in the post office building to keep the lines of communications open. My uncle, who used to live with us, was summoned to military headquarters. Every young man was mobilized to defend equipment and strategic positions in Macao.

I remember seeing very abusive slogans daubed on the walls, attacking the British and Americans as well as the Portuguese. And I remember a few months after the 1 2 3 event seeing Chinese men with red armbands controlling the traffic. The local government's authority was so weak at that time, they really did need help managing the traffic; I saw it with my own eyes. For two days, we were evacuated to a safer home, away from the Government Palace. A sort of a blockade was set up so Portuguese people, both those from Portugal and Macanese like us, couldn't get fresh food from the markets. Day-to-day supplies were completely cut off. Chinese friends used to come to our house when it was quiet, bringing baskets of things, so we still had a way of getting fresh food. It was quite frightening. We saw this as our home. I know other people who left for Hong Kong or even Portugal then, but we were quite humble people and we stayed on. Things started to get better after a few months. I think everyone realized it made sense to live together peacefully.

I stayed in Macao until I finished secondary school in 1972 and then left to go to university. It was common for people like me to go to Portugal to study, but unlike many of my friends I chose Oporto rather than Lisbon. I wanted to study economics and the Faculty of Economics in Lisbon was being disrupted by riots because of the student movement. I wasn't a very well behaved student. Macao is a conservative society and I was alone, seventeen years old, with so much going on all round me. I thought it would be more useful to enjoy myself than to study. So I spent my time learning things I couldn't learn at the university; I did carry on studying—but slowly. I got married and we had our first child in Portugal and came back to Macao in 1982.

Macao was going through a new stage of development when we returned. Just after 1974, Macao drew up its new Organic Statute and suddenly had much more autonomy than before. Everyone agreed there was a need for development here. Macao was short of people with university degrees and expertise. I joined the tourist office. At the time they only had very basic tourist statistics and were trying to develop them. In the following years, I changed jobs several times and by 1989, I was the deputy director of tourism. I had a lot of fun in those years. I was part of the first Macao Music Festival, I was on the Grand Prix committee, I was even part of the Miss Macao organization. We started lots of interesting projects.

That experience with tourism really helped me in my current job as mayor. I learned the importance of preserving Macao's heritage if we want to attract tourists. I also became keenly aware of Macao's image problem. It motivated me to start the Clean Up Macao campaign, which I launched in 1993. Of course, we always tried to be positive about these changes. In one of my first interviews after becoming mayor in 1993, someone asked me: 'Macao is a city full of problems, with the world's highest population density. It's growing so fast, there is so much building going on—how are you going to cope with it?' I said: 'I'm lucky. There are some mayors whose problems are exactly the opposite. At least my problems are all to do with Macao's growth, not its decline.' It is true that the growth process is riddled with contradictions and we need to bring the benefits of growth to most ordinary people. Now, of course, Macao is still a city with problems—but at least they are new problems.

Officially, my term of office runs until the year 2001, but there is still a lot to be defined. The Basic Law is unclear about what will happen to the municipal councils. I'm confident they will be kept, because they're doing a good job. I think if major changes were introduced here, the future government of the SAR would have some new problems to cope with. I don't think they want new problems. But that doesn't necessarily mean I personally will stay in this position. I think if I'm still sitting here in the year 2000, I will probably have been reconfirmed in the job by the Chief Executive at the time of the handover.

Even local people sometimes ask why we need a Leal Senado as well as the Legislative Assembly. Of course, the Senado isn't a Senate any more. It was founded in 1583 and until the mid-19th century it was the de facto government of Macao. Some people even say it was the first democracy in the Far East. But after that, it was crushed to our present status. The Senado is only responsible for the city of Macao, not the islands which have their own local councils. As a municipal council, we're in charge of the management of the city. We have a high degree of autonomy but, politically speaking, we have to follow the policy guidelines of the government. We can't pass laws, only regulations. If you want to leave your rubbish out, you have to do it at a certain time—that sort of thing. We don't have the power to collect taxes. We collect fees for services we provide. Even if we do collect some taxes instrumentally, we do it in the name of the government.

Even so, one could still ask why we need two levels of government. I'd say that although Macao is very small, it has all the complexities of a city state. In Macao we issue our own money, have our own air traffic rights, we're part of international organizations, we have our own judicial system, and our own laws. So the Senado deals with one level of problems and that leaves the government free to deal with another level. If we function well, the government never has to spend time worrying about sewage, hawkers, fresh food markets, or public health problems.

Within the council, most of the business is done in Cantonese now, and I speak it pretty well. It's too late for me to learn how to read and write Chinese properly. But I think it's important that communication here is good. We employ 1,500 staff, and since 1 January 1999, we have been fully localized. Of course you have to understand Macao to understand what that means. What does local mean anyway? I'm a local. I've earned the right to be called local, my family's been here for five generations. Macanese are locals. We've given our lives to Macao. The Basic Law talks about permanent residents, not about Chinese and Portuguese. Permanent residents have certain rights. We know it says only someone who is Chinese can be Chief Executive or occupy other top posts, but it says nothing about the lower-level jobs. Some people say there should be a correlation between the number of Chinese in the population and the number of Chinese in charge here. That will come eventually but we can't use that as a basis for employing or promoting people.

Of course we want to localize but we're doing it fairly. We don't promote people just because they are Chinese or Macanese or whatever. They've got to be properly qualified and have the right experience. If we didn't take that approach, the government's efficiency would really suffer. We do worry about discrimination against the Macanese in the future. We think we're part of Macao and we want the new Special Administrative Region to preserve Macao's way of life. Without identity, we're nothing. We can't compete in terms of size with any region of China. If we don't keep our specific qualities, and we don't try to develop them, we're lost.

I don't want to turn Macao into a museum. The fortresses and houses and churches are well preserved and they give Macao its background, but more important is what is inside: the people, and their culture, and the way they think. That is what makes Macao different from other places. Preserving that uniqueness shouldn't just be the responsibility of the Macanese or the Portuguese. They only have a limited influence anyway. It should be the responsibility of the Government of the Special Administrative Region and mainland China to keep those traditions alive. Macao is a city in transition. A large part of the population only moved here within fifteen years before the handover. If work is done to emphasize Macao's own way of life, they will also soon develop a sense of belonging.

As for gambling, I'm not worried about it being too dominant. It could be developed even further, especially if the companies invested more in the quality of entertainment and the style of gambling. We should introduce more entertainment related to the casino business. Look at Las Vegas. Today it's a real family destination. But people often criticize Macao for being a destination with nothing for the family. I'm not saying we should not develop other sectors too, though. We do need a better balance between gambling revenues and other revenues.

The law and order problem has been a major set-back for Macao's image, but I do believe things are getting a bit better. The problem was sparked by an unfavourable economic climate that made competition more intense and brought about some methods that are more violent than usual. Whenever there's gambling, there's some sort of negative social impact. Like everything in life, we have to find a balance. Gambling has made it possible for the government to keep taxes low, for professionals and also for companies. Basically the gambling revenues have made it possible for us to develop many infrastructure projects. There are always costs and maybe this is one of them. We have to pay a price.

Macao needs several things. It needs to develop itself as a tourist destination. A theme park might help, especially if it differs from those just across the border in China. If we had something relevant to the history of Macao, that would be OK. Also, Macao needs packaging. Even without a theme park, we've got a beautiful product. But so many people don't know about it and even people who do come here miss so much because of the lack of creative packaging. We need to make use of all the assets we already have.

I like to think of myself as a Portuguese from the East. The term Macanese makes our perspective too narrow. We need to think globally.

We've absorbed a lot of Chinese influence in our cultural background. When I speak to young Chinese from the Mainland, I sometimes find I know more about traditional Chinese culture than they do because they've been brought up in a society where those traditional values weren't appreciated.

I think at the time of the handover itself, I'll feel sad. I'm Portuguese and I always will be. It's a big change for people like me who grew up with a Portuguese background. Of course people might say: why didn't Portugal let Macao go earlier? I can't answer for my ancestors. But we have done a lot to make Macao a decent place and create the conditions for the new Special Administrative Region to be a place of opportunity.

Victor Ng Wing-lok

Building Business

Like many Chinese people of his generation, Mr Ng's childhood and education were badly disrupted by the Japanese occupation and its aftermath. Despite difficult early years, he has risen to become a prominent businessman in Macao and, as a politician, an influential spokesman for the Chinese business community in the enclave.

Mr Ng was born in a village in southern China in 1930, in the province of Guangdong. He had a limited early education in a local village school, and he recalls being part of a class of about a hundred children supervised by one teacher, sitting in rows, and studying classical Chinese texts. There were no modern textbooks available to the class at the time.

His life changed when he was seven, and Japan invaded China. Like many in southern China who were able to flee, he was sent across the border to Hong Kong and attended school there. He was only reunited with his family again in Guangzhou in 1944, the year before the war ended. His education continued in Guangzhou in a Catholic school, run by nuns. In 1946, he returned to Hong Kong to try to continue his studies there.

The stability was short-lived. His family's fortunes changed with the coming to power of the Chinese Communist Party in 1949. Mr Ng's father, concerned about his own advancing age, had saved enough money to buy a few plots of land in China as a way of providing a steady income for the next generation. But as landowners, the family suddenly found itself a political target of the new Communist regime. The family's land was confiscated and Mr Ng's parents, brothers, and sister were persecuted.

Mr Ng escaped the suffering because he was already in Hong Kong, but the collapse in funds abruptly ended his education for several years. He never managed to complete middle school. He estimates he received about eleven years of formal and informal education through childhood and adolescence, in a range of institutions.

Despite these disruptions early in life, Mr Ng did prosper and is now seen as a leading businessman and prominent politician in Macao. In 1984, when already president of the Macao Export Association, he was elected to one of the limited franchise seats for the economic sector in the Legislative Assembly. He is also an appointed member of Macao's Economic Council and serves on the board of directors of the Chinese Chamber of Commerce, a powerful focal point for the local Chinese community.

Mr Ng is also recognized by the Chinese leadership in Beijing. Appointed by the Chinese government as a member of the Basic Law drafting committee, he was later invited to be a member of the Chinese People's Political Consultative Conference (CPPCC). He also serves on the Preparatory Committee, which is the main body overseeing Macao's political transfer from a territory administered by the Portuguese, to a Special Administrative Region of China.

I STARTED MY FIRST JOB in an import-export company in Hong Kong, working as a typist clerk for two or three years, until I was offered a better job as an export manager in another company. I learned all about the export business. The company was mostly exporting to Portuguese colonies like Angola and Mozambique. After a few years, I decided it was time to develop my career and become an employer, not just an employee.

I placed an advertisement in the *South China Morning Post* newspaper, saying I was looking for a partner to set up a business exporting to Portuguese territories. The next day, I received a letter from the owner of an export company in Macao, a Portuguese man. I came to Macao in 1960 to join him and worked there until 1968.

My boss died in 1966, in the aftermath of the 1 2 3 affair in Macao. He couldn't cope with all the political turmoil and the fact the Portuguese lost so much face. He died of a heart attack. Before the 1 2 3 rebellion, the Portuguese really thought they were superior to the Chinese. Most of the Chinese weren't being treated equally, and this was an opportunity to take revenge. That's why some Portuguese or Macanese people were treated badly then. I remember seeing a few Chinese men setting on a Portuguese man in the street for no reason, and beating him up. The Chinese in Macao had suffered injustice for so many years and this was their chance to stand up for themselves and demand more rights. It changed the status of the Chinese in Macao. They did get more respect and fairer treatment after that. Of course, the Portuguese Revolution came along in 1974 and Macao developed economically, too, which also helped. But my boss was really disturbed by the rebellion. I don't think he could stand seeing Portuguese treated like that by the Chinese and it caused his heart attack.

After his death, I carried on working for his wife for two years and then left to set up my own export company. My boss had been a founder of the Macao Export Association and I'd helped him when he was setting it up. When I started my own business, I joined the Association in my own right and became a director. A few years later, the serving president died and I was elected Chairman to fill the vacancy.

In 1984, I was nominated by the Chinese community to run for a seat in the Legislative Assembly, representing the economic sector. Mr Ho Yin was the leader of the Chinese community at the time and he nominated me. It was an indirectly elected seat—if it had been directly elected, I might not have won it. I won the election because I was supported by all the members of Macao's economic sector, following the recommendation of Mr Ho Yin. That was it—I had started my political career. Since then, I've become more and more involved in politics. I feel strongly about what is just, what is right, and I've got a reputation for being outspoken and for criticizing the government when I think they deserve it.

When first elected, my main complaint against the Portuguese administration was about the level of bureaucracy and the export regulations. The bureaucratic methods used by the marine police, the quota systems, all the documentation—they're all bad for business. Business people were angry about all these measures, but at that time there wasn't anyone who dared complain. So I started to speak out. Macao is a small place. In Chinese, we have a proverb: when there are no fish in the sea, even the shrimps become big fish. I wasn't an outstanding leader, I was just a shrimp, but this was a pond with no fish in it. No one was prepared to criticize the government and speak for the people. I might have been one of the first Chinese people to criticize the administration so publicly.

Of course, we've got to be realistic. We cannot say Macao hasn't developed at all. It has. The administration has been modernized and the basic infrastructure has been improved, with the building of the airport, the new bridges, and the new pier and so on. But the administration hasn't diversified the economy. On the one hand, they've developed the infrastructure, but on the other they're far too conservative about attracting foreign investment. They haven't really cut down on the bureaucracy either. Maybe it's just the Portuguese way. We've developed the infrastructure, the hardware, but there's no software: no foreign investors, no foreign capital.

I've tried many times in meetings with them to persuade them to be more open, to open the door as wide as they can. But still, they do nothing. Look at China's example. Why have the open door policies been such a success in the last twenty years? Because they're open, they're flexible, they know what the investor wants. But Macao's attitude to investors is this: 'You can come if you want to, but if you don't want to, that's fine by us.'

My company is a holding company with two factories in Macao and one in China. Our main business is exporting garments to European and American markets. Until five or six years ago, we used to have a triangular business between Macao, China, and Portugal, shipping merchandise from China to Portugal with payment and arrangements in Macao. Now Portugal and China have both developed, so we no longer have a role to play: goods go directly between China and Portugal. Our business has suffered because of that, but we're looking ahead. If there's a better business environment in Macao, we can still thrive.

At the time of the handover, many issues remain outstanding. The localization of the civil service, of the legal system, of the Chinese language—there is still a distance to complete. Even the stationing of troops has become a point of contention. We Chinese should accept that these issues won't be resolved in negotiations with the Portuguese, and must wait until after the handover. There's no mention in the Basic Law of the stationing of troops, but that doesn't mean it can or cannot be done. It's up to the central government in Beijing, as defending Macao is their responsibility. Bringing troops here is a way of showing China's sovereignty over Macao. Internal security isn't the job of the troops, it's the job of the internal security forces. But Macao has such a big law and order problem, people aren't sure the internal security forces can cope. So I think stationing mainland troops here shows sovereignty, but is also the Central Government's way of showing support in tackling the issue of law and order.

The security problem has developed because the casino operators don't run the business according to the law. The gambling law says the casino cannot be sub-let or sub-contracted in any way—but they do exactly that. There are so many gambling halls that aren't really owned by the official operators. Because these gambling halls are owned by different people, they're vying to protect their own vested interests. They invite the triads to be their partners, or ask them to give them protection. The money at stake is so great that once the triads get involved, it soon leads to in-fighting. This is the root of the triad problem. I've consulted a lot of lawyers about this issue and they say the law has been violated. I think the government is well aware of what is happening, but is deliberately turning a blind eye. It's a political decision.

Macao's Legislative Assembly is quite different from Hong Kong's. The Basic Law states that for the next ten years, seven members will still be nominated by the administration. The only change is an increase in directly elected seats. Some people say they want to revise the Basic Law to increase the pace of democracy, but I don't think that's a good idea, it's such a complicated process. Democracy is the long-term goal but it may take much, much longer to come to Macao than to Hong Kong.

Am I looking forward to the handover? Every Chinese resident of Macao is looking forward to it and, yes, that means me too. It's a great opportunity for the Chinese. We'll run Macao much better. I don't mean the Portuguese administration was a complete disaster. They have done some good things, but Macao is a Chinese territory and it's natural that it should be run by Chinese people. Look at the language issue! Once the administration is mostly Chinese, it can start to use Chinese and life will be much easier. A sense of Macao's unique culture is important, but only to attract tourists. For four hundred years, the Portuguese didn't promote their language in the community. They never had a fixed policy for promoting Portuguese; and local people saw the language as narrow, something they could only use in government offices.

Macao could still be a centre for studying Portuguese, but we cannot force people to use Portuguese in the community. If people want to use

Chinese, you cannot make them use Portuguese. In ten or twenty years time, the main language used here will be Chinese. Macanese people will gradually leave anyway, the younger generation will go overseas to study and won't come back. The number of Portuguese and Macanese will be fewer and fewer. And if they do stay, they'll have to integrate and accept Chinese culture.

We Chinese know what we want for Macao better than the Portuguese do. We want Macao to be economically more developed and we want the fruits of Macao to be shared fairly. Take the gambling industry, for example. The operators have done a lot of good things for Macao, I don't deny that, but the tax they're paying the government is far too small. In the future, when we review the licensing system, I think we have to change that and make sure more profits go to the people of Macao, and not into the pockets of private operators.

I accept the casinos should be a major part of the local economy— but that doesn't mean we should ignore other opportunities for development. We need to diversify. I think we have the potential to make Macao a centre of garment supply. I don't mean all garments have to be made in Macao, but from the year 2005, when all the export quotas for garments are abolished, I think we could export garments made on the Mainland, but with the supply, ordering, and quality control done here by Macao merchants. You want to buy garments? Come to Macao—we provide a full serivce. And the garments wouldn't just be from China, but from South Korea, Thailand, Vietnam, and so on. An idea like that needs open policies from the government. If they're as conservative as they are now, we cannot do this.

We have to look at the past as history and look forward. You can't dwell on the problems of the past. If you focus on the bad things, you don't have any room to develop. It's politics. We have to look to the future. If I were a revolutionary, what would I achieve? We have to be realistic. We have to put the country first and the personal memories second. My family suffered greatly because of the political turmoil in China. I haven't forgotten that, but in my role I can't mix the personal and the political.

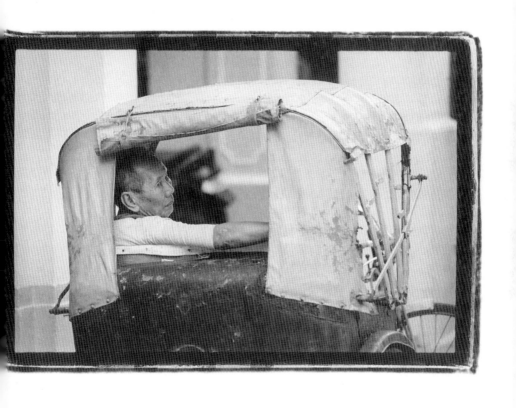

Ng Kuok-cheong

Fighting for Democracy

Ng Kuok-cheong is an unusual Legislator—and in many ways he is the odd man out in the Legislative Assembly. He has no distinguished family lineage, and no powerful family connections. He fits neither into the leftist, pro-Beijing camp nor the pro-Macanese/Portuguese one, but often finds himself instead a lone voice of protest in the Assembly, caught between the two.

He came to Macao soon after his birth, a Mainland immigrant. Details of his parentage remain a mystery. His passion for politics has dominated his life, first as a fervent anti-colonialist in the late 1970s, joining demonstrations in Hong Kong at that time, later campaigning in support of the pro-democracy demonstrations held by mainland students in Tiananmen Square in 1989. More recently he has poured his energies into championing greater democracy in Macao, calling for more government accountability, universal suffrage, and economic restructuring. He is a leading member of the New Macau Association, a pro-democracy campaigning group. Ng Kuok-cheong is the only member of his party to win a seat in the Legislative Assembly, to which he was directly elected.

I DON'T KNOW WHERE I WAS BORN. I don't know who my parents were. I was adopted by a woman when I was a baby. I might have been born in Macao and sold by my parents here, or born in a village in China and then sold later. I've had some hints recently about who my parents were, but I don't really want to find out more.

When I was a child, I always knew I was adopted. They didn't actually tell me, but there were plenty of clues. The woman who adopted me told me I was her younger brother. Her father and mother had already died. But I saw their tombstone and the dates showed they'd both died before I was born. There were many other signs, too. In any case, I grew up with this woman, calling her my sister. I was born about 1957 or 1958, a time when many of the villages were very poor, and a lot of people came to Macao as refugees. They couldn't take their sons and daughters with them so they sold them.

My sister never married and she was an only child. Chinese people always want to carry on their family, and adopting me was a way of doing that. My sister had two or three properties and earned an income from renting them out. Her father had been a middle-class landholder in China, and at the time of the civil war they'd left the Mainland and come to Macao and invested their money here.

I would describe my childhood as one of hard work. I knew I wasn't my sister's real brother so I tried to work hard as a way of earning my place in the family. Only about 5 per cent of people then attended Portuguese language schools in Macao—so like most other people, I went to a Chinese school.

I think there have been a series of major changes in Macao in my lifetime. The first was in 1966, when the Chinese rebelled against the Portuguese governor—successfully. From then on, I felt the police lost their authority in the street and Chinese people won their dignity. I was only a child at the time and I attended a Catholic school, which of course didn't support the rebellion, so I wasn't directly involved with what was happening. But I remember hearing the revolutionary songs about Chairman Mao sung in the street and on the radio. They broadcast songs by lower class people, not the rich. This was very significant. Every year on 1 October, Chinese National Day, there was a tremendous celebration—and the celebrations on the tenth of the tenth [the date 'ten ten' or 'tenth of the tenth' refers to the anniversary of the founding of the Republic of China on 10 October 1911] suddenly stopped. That was one change.

My sister was a small landlord, so she didn't support the rebellion. Some people, friends of my sister, said they'd leave Macao because of the unrest and some local Portuguese said the same thing. They felt the working-class people gained a kind of strength, that they became a powerful force on the street, and weren't very polite to higher classes. My sister's friends felt threatened by that. Some of the Catholic schools were attacked by the rebels, and one of the Christian schools changed its colours and became a Left camp school at that time. My school wasn't close to the city centre and it stayed calm. The whole incident did bring some changes. For one thing, policemen started being more polite to ordinary Chinese people after that.

The second big change was Portugal's revolution in 1974. We didn't realize its importance at the time, but really it was a very important development. That's when Portuguese troops were withdrawn from Macao, and when we established the Legislative Assembly, and the local Portuguese were given the chance to take part in government.

The third change was the open door policy adopted by Deng Xiaoping in the late 1970s. People flocked to Macao from the mainland, and manufacturing industries here grew rapidly. At the same time the gambling industry flourished, too. That was partly due to Hong Kong people; they enjoyed rapid growth, started to get rich, and came to Macao to gamble. So then in the 1980s, Macao saw quick development. New buildings went up everywhere.

When I'd almost finished school, my sister became seriously ill, and was in hospital for about three years before she died. When she died, the flats passed to her distant relatives. They gave me income on one flat, but took the money from the others. It was about 1978 and I went to Hong Kong to study at Chinese University for four years, reading economics and social sciences. I applied for every scholarship I could and worked as a tutor for local children to earn my keep. I lived in university accommodation, and managed to make ends meet. When I graduated, I wanted to stay on in Hong Kong. I didn't have any friends or relatives in Macao so it didn't matter to me where I was, but I only had a Macao ID card; although I applied to stay on in Hong Kong, I was refused. It was a sensitive time. Hong Kong was starting negotiations on its future and they didn't want pro-China activists there. During university life, I'd been very active in the student movement, protesting against the government and police, and arguing that Hong Kong and Macao belonged to China and not to colonialists, so that may have been a factor.

When I came back to Macao, my first priority was to get a job. I joined the Bank of China as a junior. I felt Macao was being offered a great opportunity with the political change ahead. I could see that the Left camp, in line with China's open door policy, was eager to cooperate with the government in Macao and with the other social agencies, including the Catholic groups and the local Portuguese. They also opened the door to the intellectuals in Macao.

I took part in all the activities organized by the Bank, which was a member of the pro-China camp, and started volunteering at a Catholic youth centre as the editor of a magazine there. There was also a society formed by local intellectuals. In Macao, very few people were included in the so-called intellectual class. They usually came from mainland Chinese cities; Guangzhou, Beijing, Shanghai. Some were born in Macao, but very few. They became very active because they also felt Macao had a very promising future and that Chinese people here could take part in planning that future. Most of the members agreed Macao needed economic change. We could see that local manufacturing, based on cheap labour, wouldn't survive the 1980s so the first priority was for economic restructuring. Also, we wanted to prepare for political change because the Portuguese government would soon leave Macao. We wanted more study on these two critical areas.

In 1989, when the pro-democracy campaign was underway in Beijing, we organized a series of activities in response. Even after the June 4th crackdown, we tried to carry on supporting the students. The Bank was very reluctant about letting me stay with them so in the end I had to leave. They asked me to stop my political activities, and I wouldn't do that. The pro-democracy movement was a real opportunity for China to go in a new political direction. We thought it was a chance for change and of course we supported that. We knew the seeds of political revolution could mean a different outcome. Of course I was very upset about the crackdown, but we were calm too. We were well aware it was one of the possible outcomes.

After job hunting for several months, I found a new job in Caritas, the Macao division of the international Christian charity that has many local welfare and educational programmes. I worked in the accounts department. We formed our society, the New Macau Association, in 1989, as a way of keeping up political pressure on China and the Macao government and we took part in the 1992 elections to the Legislative Assembly. I'm one of the Association's four vice chairmen and I won our only seat out of the eight directly elected seats. We were determined to continue proposing new laws and strategies for Macao, even though, with only one seat in the Legislative Assembly, we were always doomed to fail. It's one against twenty, but we keep trying. No one wants to cooperate with us; that would make them unpopular with either the Left camp or the Portuguese members.

It's hard to explain our political strategy for Macao in a few words. The main points we emphasize are that the Macao government should be supervised by Macao people, and that Chinese people on the Mainland should have the right to criticize Beijing. We think anyone representing the public in political institutions should have a manifesto, outlining their strategies in different areas. We're calling for Macao to have a directly elected Legislative Assembly, and a directly elected governor, and we want to fight for this after 1999.

Most of the local representatives involved in drafting the Basic Law came from the left camp and all they wanted to do was identify what Beijing wanted and implement that. There was no independent thinking, they didn't have ideas of their own. Of course we wanted the process to be more democratic and for the drafters to be elected by Macao's public, but that never happened. Even Hong Kong didn't manage to go that far. We suggested various policies for economic change but the government wasn't interested. All they wanted to do was keep the economic figures healthy until 1999, to keep up their tax revenues—and that meant leaving gambling as the main industry. Their priority was to preserve Portuguese culture in Macao and they needed money to do that. They wanted to leave hard-core infrastructure for Macao—the new airport, new hospitals, new bridges—and they needed money for that. Before 1987 there was only one health centre in Macao—now every area has its own. They also set up a social security fund in Macao.

The government had already decided what they wanted to achieve before the handover and they had to keep revenues high to do that. So they turned a blind eye to the way the gambling industry operates, and let a lot of money flow into the hands of the criminal gangs. They didn't care—all they cared about was their own tax revenue. That's how the law and order problem developed in Macao. The government also let the gambling industry set up a sub-contracting system, so money started being siphoned from the legitimate business to feed the criminal organizations. These gangs started to work with elements in the Macao police to fend off Hong Kong gangs who wanted a share of the profits— and that's led to gangland civil war. Because the Macao gangs have a close relationship with the police and some government officials, the authorities can't get control of the situation. The Macao government doesn't feel strongly about it anyway. They're trying to get the most they can out of Macao before 1999. The gangs are doing the same. After 1999, the gangs will quieten down and try to get along peacefully with the new chief executive. But until 1999, they won't give the Portuguese face.

Our fear is that the new chief executive will accept the criminals' offer to make peace with them. If he does, the gangs will still exist and carry on getting richer and more powerful, and still be able to bribe government officers. Then, if another power struggle breaks out, the new government won't be able to stop them either. The only option left would be to involve the Chinese army, and that would be dreadful. The prospect of China's army being stationed in Macao is a problem in itself. Under the one country, two systems formula, the central government in Beijing has the right to station troops in Hong Kong and Macao if necessary. Hong Kong's Basic Law has articles that say the army must obey local laws and can't interfere in local affairs. In a crisis, the chief executive can ask the central government to deploy the troops to help keep law and order.

In Macao, we asked for the same article in our Basic Law but the drafters said it wasn't necessary, because there were no plans to station troops here. So they refused to include any provisions, despite our plea. I'm not saying the Central Government won't tell the army to obey local laws, they'll do their best because they know the world will be watching. But what about in fifty years' time? First they say there won't be troops in Macao. Then they break their promise and say there will be troops here. Now they're telling us the troops will obey local laws— but in the future? We've been left with no legal protection about how they behave. They're also implying the troops are coming to Macao to solve our internal security problem—which of course is not their job under the Basic Law. The motivation is all wrong.

Democratic reform is also key. Unlike Hong Kong, Macao doesn't have a strong lobby calling for more democracy. There's only a tiny middle class. There are lots of people earning middle-class incomes, but they're pseudo middle class because they're managers in the casinos or very junior civil servants. They do semi-skilled work but earn a middle-class income. The real middle class, which might champion democratic reform, is tiny. The other difference with Hong Kong is the impact of the rebellions in the 1960s. In Hong Kong, the British government defeated the rebels, but afterwards they tried to absorb some of them into the government structure as a way of keeping social stability. In Macao, the opposite happened. The rebellion was successful but it was based on a concept of cultural revolution, which saw the state as the apparatus of repression. So after the rebellion, the leftists didn't ask for political reform, and didn't ask for a voice in the government. Instead they concentrated on working at the grassroots level where they're still very successful.

We want to show people that society can be changed through the existing political structure, through the ballot box. There's a new immigrant generation here, which came to Macao in the 1980s and will come of age about 2005. They're not connected to the left camp, but they make up about 40 per cent of the population now, and they might want to play a part in shaping Macao's future. If they do, we'll have a chance of pressing on with democratic reform.

All we can really do in the meantime is call for reform. The Legislative Assembly is so conservative that when we do anything, we're breaking new ground. If we stand up and criticize the government in the Assembly, the mere fact we're doing that is progress in itself, because until recently, no one did that.

General Vasco Joaquim Rocha Vieira

The Last Governor of Macao

General Vieira was sworn in as Macao's last governor on April 23rd 1991. It marked his return to the enclave. Previously, he had served as the Chief of Staff of the Independent Territorial Command of Macau in 1973 to 1974 and Under Secretary for Public Works and Communications of the Macao Government in 1974 to 1975. It is from this first stay that he begins his account.

WHEN I ARRIVED IN MACAO in September 1973, it was a totally different place from Macao today. In those days, people didn't have as much concrete experience of China as they do now. The Far East was an almost mythical place, far away from Europe. I'd lived in Africa and Europe and was really curious about China and exploring the culture here. China was still closed to the world then, and we felt there was a great culture hidden behind the curtain, difficult to access. Macao was a small city but it impressed me from the start. I could sense it was something special. The smells, shapes, rhythms, and music of the place were totally new to me but as well as being oriental, it was also a very friendly environment, and I found it easy to feel at home here. Macao wasn't very developed economically at that stage. It reminded me of the sort of places I'd read about, still quite traditional societies with little opportunity to develop and only limited communication with other places.

Macao changed following the revolution in Portugal. The revolution gave Macao new chances. In Portugal, we were able to approve a new political status for Macao, giving the enclave a large degree of autonomy, and establishing a representative Legislative Assembly with direct elections and more involvement of the local population in decision making. We established a clear system with democratic roots and independent courts. It was a new departure for Macao and reinforced its identity. The new regime in Portugal also made it possible to start negotiations with China to re-establish diplomatic relations between the two countries, which occurred in 1979. This was of crucial importance and made it possible for Macao's authorities to start speaking directly to the Chinese. It also gave Macao the chance to benefit from the opening up of China, from 1979 onwards.

Even the mentality of the local people changed at that time. When people enjoy freedom and liberty, and are able to express their wishes more freely, it changes the whole relationship between the administration and the population. In those days, most of the Chinese community lived a bit apart from the administration. Only a small section, those involved in commerce and business, had direct contact with us. The relationship between the administration and the Chinese community was friendly, but conducted mainly through the representatives of various associations and organisations partly because the population of Macao was very transient. Many people came here from mainland China and passed through to settle in other countries. These people didn't speak Portuguese and weren't used to Western ways, and they tended to live separately. But we had a good relationship with those who did relate with us regularly and I made a lot of Chinese friends. Of course there was also the Macanese community. Within the Portuguese and Macanese circles, everyone knew everyone else. If someone arrived from Portugal, everyone knew there was a newcomer. You'd go for a walk in the street and people would nudge each other and say: 'That's him—he's the new arrival. He's come to do such and such.' People got together to play tennis or mah-jong, or to hold dinner parties.

I learnt a lot in those years, from my Chinese friends and from the local Macanese residents, which helped me when I came back fifteen years later as Governor. I learnt things then which, if I'd arrived for the first time, I could never have known. As Governor, it is more difficult even to make good friends and to discuss certain issues. I had kept in touch with a lot of my friends in Macao so, when I returned, I had a circle of old friends. That has helped me a great deal. Sometimes, when

I have to make a decision about something, I talk the matter through with these old friends, in a way that wouldn't be possible if I hadn't known them for so many years. Sometimes they warn me to look out for some factor I hadn't considered, or they offer solutions. Probably the most important thing I learnt in those years was that my own viewpoint isn't necessarily the right or only one. Sometimes there isn't just one single version of the truth. One has to be able to put oneself in other people's shoes and take into account the fact that other people have different perceptions, different cultures, a different sensibility. Before making a final decision, one has to take into account other equally valid opinions. It is a question of respecting differences. Here, the majority of the population has a different culture from my own—I can't impose my ideas without trying to get a better understanding of their perspective.

Macao has changed a lot since that time. It is more modern and more competitive, but it still keeps its own sense of identity. The modernization process has taken place all over the world, of course, that's to be expected. To me as Governor, the most noticeable difference was the change in perception of local people. There is a new sense of uncertainty, a new awareness that people are facing a time of transition and don't know what is going to happen after the return to China. One of my most crucial tasks as Governor was to address this uncertainty— to establish the structures that would allow people to feel confident about the future, to set up the legal system, to compromise with China, and to develop a spirit of cooperation so people could be sure things would go according to the Joint Declaration.

Those who have to take decisions must read, listen, and talk to people. As Governor, I spend most of my time talking to local people. In the mornings, I work with my secretaries and other members of the administration. In the afternoon, or even at lunchtime, I talk to people. I do my best to see everyone who wants to talk to me. Macao is a small city and of course it is very useful to walk on the streets and listen to people talk. I also often speak to the media because wherever I go I am expected to answer questions, but it's more important to speak directly to small groups of people who can pass on the message. I address a lot of clubs and associations, go to local festivals and openings, and walk around the streets without attracting a lot of publicity but with my

interpreters to help me communicate. That's why I realize that some of
the criticism one reads in the newspapers doesn't match the feeling I
get from local people, and most of the time, what the local people are
saying is closer to what I think myself than some media reports.

Take the economy, for example. People with whom I speak
understand that Macao was partially struck by the general economic
climate. They compare Macao to other places in the region and say
we're not doing too badly. They recognize that Asia has been hit by the
financial crisis and they offer positive solutions, not just criticisms.
People are indeed very reasonable. From the starting point of a small
and highly dependent economy, a few years ago Macao entered an
important modernization cycle, and gained the appropriate tools needed
to strengthen its own autonomy and to become a useful platform in the
region. When it was prepared to take advantage of new opportunities,
then came the Asian crisis. Nevertheless, people recognize that we in
Macao did not commit the mistakes other did, and that other
neighbouring economies are facing more problems with the crisis than
we have in Macao.

Our intention was always to establish good cooperation with China.
I see that as best for the population, and as the best way of maintaining
confidence. That spirit of cooperation also makes it easier for us to
convince China that our ideas and policies are the right ones for Macao.
The Macao government knows better than the Chinese authorities what
Macao—the second system in the one country, two systems formula—
is all about, because we live here. We state our position firmly—but in
open discussion. I think that's best, and it works.

We see the handover as a point on a continuing line, rather than as
the end of one era and the beginning of a completely new one for Macao.
In political terms, we can't separate the meaning of the handover
ceremony from the way we've dealt with the transition with China. We
want the meaning of the ceremony to be in accordance with that spirit
of cooperation. Macao has a historical memory, it has a political system
in place, it is an example of the understanding between two different
cultures and civilisations. We've done everything possible to preserve
the identity of Macao, so that it can continue to fulfil its role as a bridge
between East and West, and continue to be a unique place in terms of
cultural understanding. The handover is only a change of administration.

Macao itself will continue as it was. It's essential that Macao retain the sense of its own special character, while becoming part of China. I don't mean I want to preserve Portugal's influence here beyond 1999. I want the present character of Macao to continue in the future. Hong Kong's last governor, Chris Patten, used to say Macao made him nostalgic for Europe. The Chinese authorities understand that Macao must preserve that special characteristic.

What have I learnt in these years as Governor? It has confirmed to me that life is very complex, that even a small place like Macao is very complex. To achieve results, one has to work with determination, be able to understand other people, and be able to motivate other people to work with you. One has to have good judgement—but also not work alone. Macao is facing a great transition, and we have to find the right path. I will always remember the years I spent here. I'll remember this unique process of transition, something for which we have no previous experience, no reference points. I hope we've done enough to make sure Macao goes forward to success.

History and Heritage

The past century has transformed Macao, both economically and socially. Some people criticize the enclave for not having changed enough, particularly in economic terms in comparison with its dynamic neighbour Hong Kong, whose skyline reinvents itself every decade.

For Macao's older generation, those now in their seventies and eighties who have borne witness to much of the past century, the enclave today is a foreign land compared with the home of their childhood. In this section, five older people speak wistfully about their memories of the sleepy, peaceful Macao before World War II. They paint a picture of a place with two distinct communities, one Chinese and one Macanese, imbued with a sense of tradition and family heritage.

For the Macanese and Portuguese, the neighbourhoods were defined by the Catholic churches, with a different community growing around each of the major parishes, taking their identities from them. Many lived in large mansions with lengthy verandas and individual gardens, set along broad leafy avenues. Most households were supported by several servants. Parties and cultural activities were a part of daily life.

The Chinese community lived in far more cramped conditions, crowded closely together in town houses along narrow streets, which began the century as mud roads but were gradually paved with large cobbled stones, Portuguese style, as Macao developed. These were commercial districts as well as residential ones, the centre of Macao's daily business life, with noisy street markets and shops.

World War II brought an end to those days of peace and plenty. Firstly, in 1937, Japan invaded mainland China. A few years later, in 1941, Hong Kong fell to advancing Japanese forces. Macao itself, like Portugal, was officially neutral during the war and escaped the levels of hardship and suppression documented in both Hong Kong and China, but the impact of the war was still felt in a different way. Refugees poured into Macao, fleeing from Japanese occupation. The population swelled dramatically to reach about half a million, which for Macao in wartime conditions was close to bursting point.

Some people who were in Macao during those war years describe regular killings by the Japanese authorities. Others say they experienced a relatively stable and peaceful period. The greatest hardship was hunger. Those who had special access to food supplies, or who were wealthy, seem to have escaped unscathed. There was one particularly acute period when rice supplies, accessed from China by negotiating with the Japanese forces, were badly disrupted. Starvation became so prevalent that dead bodies were frequently seen by the roadside, where victims had collapsed from weakness.

The first section touches on the political importance of the 1 2 3 incident, which occurred more than twenty years after the end of the war. This section includes first-person accounts of the acute civil unrest in December 1966, which polarized society and had a long-term impact on the authority of the Portuguese administration.

The politically explosive confrontation between the Portuguese-led administration and Macao-based Chinese leaders, many of whom were gripped by the fervour of the Cultural Revolution taking place just across the border, changed the relationship between the two communities forever. Here Mr Chui Tak-kei, a senior member of the Chinese community and an advisor to the local Chinese leader, Ho Yin, during his negotiations with the governor of the day, adds the Chinese perspective.

Mr Arnaldo de Oliveira Sales

A Gentle War

Mr Arnaldo de Oliveira Sales is one of Macao's best known sons. His record of public service in Hong Kong spans several decades, including being the elected Chairman of Hong Kong's Urban Council for four terms from 1973 to 1981. The list of his honorary and active posts—and of his awards—runs to several pages and covers the worlds of business, community, and youth work.

His most outstanding achievements involve his passion for sport and his belief that sporting opportunities should be available to everyone. He championed the development of public sports facilities in Hong Kong and made neighbourhood play areas available for the young, and rest gardens for the elderly. He has supported the growth of amateur sports associations and events at all levels, from local clubs and national sports associations in Hong Kong to regional and international competitions. He has worked prominently in the Olympic Council of Asia and the Commonwealth Games Federation, of which he was Chairman from 1990 to 1994.

His family, well established and influential in Portuguese society in the Far East, settled in Macao five generations ago. Both his parents were born in the enclave. His father's ancestry is an unbroken Macao line from Lisbon. Three of his mother's grandparents were born in Portugal. Her maternal family, too, had been heavily involved in Macao affairs, especially in the nineteenth century. Many of her ancestors were military officers who came to Macao and married into powerful, prominent local families. Dr Sales Marques, the de facto Mayor of Macao, is his cousin, as is the husband of Mrs Anabela Ritchie, President of the Legislative Assembly.

Mr de Sales was born in the foreign concession of Shamian in Guangzhou. He remembers early childhood as a pleasant colonial one, with morning outings to the park, accompanied by his Chinese amah, and grand buildings with shady verandas. His great-grandfather had served as the French Consul there in the mid-nineteenth century, one of a number of postings along China's East Coast. Mr de Sales' grandfather had set up business ventures in China, and his father joined him as a young man to help out.

The family moved to Hong Kong when Mr de Sales was nine and, while his mother tongue is Portuguese, English became his working language. Although his family remained in Hong Kong, Mr de Sales was sent back to Macao several times for schooling there—his father was keen for him to have a Portuguese as well as an English education. The first time he was really based in Macao was during World War II, and he begins his account with his memories of wartime.

I WAS ABOUT TWENTY-ONE when the war broke out. The whole family had been evacuated to Macao by the Portuguese government, when Hong Kong came under Japanese occupation, and they stayed in Macao throughout the war. I didn't join them straightaway. I stayed behind in Hong Kong for a while, to see if there was a chance of salvaging clothing or anything else from our house, which had been requisitioned by the Japanese. The family had left in such a hurry, they'd only been able to take what they could carry in their hands. I had been issued with all sorts of passes from the Japanese Foreign Office and went along to the house to try to get in. I made it as far as the garden when the Japanese guards rushed up and stuck a bayonet at my stomach and shouted: 'Stop! You can't go beyond this point!'

When I did arrive in Macao, I worked half a day with the refugees from Hong Kong, with the Executive Committee for Refugees, and the other half a day I carried on with my studies. A lot of people who came from Hong Kong had to stay in refugee centres. The government requisitioned some of the larger institutions, like the Bela Vista and Club Macao. We didn't stay in one of these centres, we had our own accommodation. When we first arrived, we stayed with relatives, then we leased our own place. There were other Hong Kong families who did the same. The houses weren't luxurious but one made do.

I went to a few extramural classes at the Jesuit school in Macao, run by the Irish Jesuits and set up to look after the Hong Kong Portuguese boys. I also went back to the Portuguese Seminary school, where I'd been a pupil after my studies in Hong Kong, to take part in discussion groups and other studies. I read and studied a lot during the war, and played bridge as a pastime and in tournaments regularly. I was also courting my wife, also from a Hong Kong Portuguese family evacuated to Macao. We got married after the end of the war.

The refugee work was run from the Santa Casa da Misericórdia, an institution founded in Portugal over 500 years ago, and established in Macao shortly after the Portuguese arrived. An early example of European community service, the Santa Casa helped the disadvantaged. The institution still exists in Macao, located right in the city square. During the war, the government set up a committee headed by a few of the older citizens, and some of us younger people worked there in various ways. For me, as for many, the added incentive was to have something useful to do, but working there also meant one was classified as a semi-civil servant and thus had access to rations.

Rationing was a great preoccupation during the war years. Once a month, each of the families went along to the collection point and was given a ration of so much flour, so much rice, so many catties of sugar, and so forth, according to the size of the family. I was a bachelor living with the family.When we were evacuated to Macao, two of our old loyal amahs went with us. They didn't want to go back to their villages in China and they were close to some of my sisters, having looked after them since they were little. So they came with us and therefore we had domestic help during those war years. Nevertheless, like everyone else, we had to be careful with our rations, though one could buy freely in the market. In a sense the rations were a reassuring stand-by, but they weren't the only food one could get.

Life wasn't altogether unpleasant. We had friends and relatives in Macao. I had plenty of volunteer work and studying to do. We had to make our own entertainment, and made do as best we could. There was a bridge club, which the British Consul and the Macao government had set up, so I played bridge in my spare time, and took part in some amateur dramatics, putting on stage plays for the local and the refugee community.

Those of us who'd been evacuated to Macao received subsidies from the Macao government and were given a monthly allowance. For British subjects, the account was rendered by Lisbon, and I think London in turn refunded them the costs. As with some other refugee families, we were lucky because we had remittances as well, a modest quarterly sum sent to us from our private sources in London. In Macao's environment, that went some way because foreign currencies had value. Although we didn't have an easy time, the government did try their best under heavy constraints, and the Hong Kong refugees were grateful.

Macao was a refugee haven. The population suddenly grew tremendously. Before the war, it was only about 70,000 but during the war it swelled to about half a million. We lived pretty much in isolation from the outside world. Life went on, the civil service worked, and the machinery of government went on under beneficent leadership in difficult conditions. But we never forgot that the Japanese were just outside Macao, occupying southern China. They exerted pressure on Macao through controlling food, letting in food supplies in exchange for whatever they wanted to take—first money, then later metal, or anything else they needed.

News spread quickly in Macao. Many people listened to the radio, and other news travelled by word of mouth. There were local newspapers in Portuguese and Chinese. An English-language newspaper called the *Macao Tribune*, a weekly published on Sunday, was also set up during the war. I still have copies at home. A few professional journalists from Hong Kong, who were refugees in Macao, ran the paper. I remember the British Consul used to say it was the only English-language newspaper between Chongqing and Hawaii.

I had a particular interest in this newspaper because I was asked to write a weekly column on bridge, which was my hobby. I wasn't paid, we were all amateurs. I made a few other free contributions, too, at various times. Things like that helped to keep life going. Every Sunday morning we used to wait avidly to read the *Macao Tribune* because it carried a lot of news about Macao. I don't remember it being political at all. It was basically intended to keep up morale.

There were also efforts to bring the two communities together; the refugees from Hong Kong and the resident community. We had friendly sporting engagements, for instance. We'd play each other at badminton, football, hockey, or tennis, and also bridge and canasta, so people did mix socially. But generally people led a quiet life. There was very little crime in Macao during the war years. One could go about freely at any time of day and night.

The main focus was on keeping up general morale. Some people were living in very tight conditions. There wasn't money to burn—just the contrary. The Macao government looked after the refugees from Hong Kong as well as they could, but the pressure of numbers was great. So a brave attempt was made on all sides to lead a normal life.

Macao has many churches, and the Portuguese arrivals from Hong Kong were mostly a church-going Catholic community. A few of the dedicated Italian priests in Hong Kong came along to Macao, too, and a church was assigned to them in Macao—Saint Augustine Church, the chapel of the original Augustinian monastery, now long gone. So there were many traditional celebrations in Macao, in keeping with the life in Portugal, which Macao still pretty much continues. There was considerable hardship, but everyone knew there was light at the end of the tunnel, and looked forward to getting back to Hong Kong eventually.

Those years in Macao taught me a few important lessons. For one thing, I realized it was possible to do without a lot of things in life. I remember losing my sense of property, the idea that one needs a lot of physical things to live, clothes and so on. When we came back to Hong Kong, we had to rebuild our lives from scratch because everything had been looted. The district we had been living in had been turned over to Japanese officers for use as their accommodations, off-limits to the troops.

I remember we were friendly with a number of the British officers, sent out to Hong Kong from London at the end of the war to help with Hong Kong's rehabilitation. We entertained them in our home and they invited us back to the Officers' Mess. On one occasion, we went to the Mess for dinner and my mother saw several items of our own furniture there, a few of the pieces my mother had brought to Hong Kong from Guangzhou. Now, of course, they'd be collectors' items. They were very happy to help us get them back by applying to the custodian of enemy property.

In recent years, my trips to Macao have been flying visits because of my commitments in Hong Kong, but Macao seems to have prepared well for the transition to China. They've been refurbishing the churches, and they have built new museums, and a magnificent cultural centre. And of course they've set up the new far-seeing university. Macao's charming hallmark is still its cobblestones. They're just like the streets of a European provincial city of the last century. How much of all that will survive, one doesn't know, but I hope Macao will thrive on its essential differences, cultural and historical.

In some ways, the handover is much the same as for Hong Kong, but there are key differences. Hong Kong is a great financial and commercial centre, and English is the lingua franca of commerce, just as Portuguese was the lingua franca here in the East a few centuries ago. That's the way history evolves. We're living through a fascinating historical period, and I'm particularly interested because I was so heavily involved in the rebuilding of Hong Kong after the war and the development of so many aspects of Hong Kong life through the Urban Council and other civic institutions.

After the war, Macao progressively sank into being a quiet, pleasant city and it only started to emerge strongly again in the last twenty or thirty years. Now, of course, it holds its own in the same way Hong Kong does. I can remember when Hong Kong was a backwater compared with Shanghai, but Hong Kong emerged dynamically and took great advantage of the circumstances and its industries developed by leaps and bounds. Now Hong Kong has gone full circle, and it's back as a service centre. Commerce doesn't play the part in Hong Kong now that it did before.

There's another key difference too. Macao has long recognized pragmatically that it was part of China. Portugal didn't conquer the territory, Macao was granted to Portugal at the request of the empire because Portugal had rendered service to this part of China in dealing with the pirates. Portugal has the high moral ground because after the Portuguese revolution in 1974, the Portuguese government gave away all its overseas territories and they even offered to give Macao back to China as well. So, with that as the starting point, the return of Macao to China isn't as difficult emotionally for Portugal as Hong Kong's return was for Britain.

The Portuguese administration has worked hard in Macao and they're leaving it with a lot of Portuguese features and cultural institutions. I think the administration in Macao, and this last Governor, have come up trumps with what they've been doing. Macao hasn't had a garrison for years, the place isn't armed, so it's not a transfer of sovereignty but more a handing over of an administration with dignity and flying colours.

Photograph courtesy of the Historical Archives of Macau

Sir Roger Lobo

Wartime Resistance

S ir Roger Lobo was born in Macao in 1923. Like Mr Arnaldo de
Oliveira Sales, he is a successful son of the enclave who left for Hong
Kong as a young man and developed his career there. He rose in Hong
Kong to become an influential public servant, businessman, and well-
respected political advisor.

His individual achievements and posts are too many to list, but key
involvements included the head of the Hong Kong government's media
department and a decade with the Independent Commission Against
Corruption (ICAC) in Hong Kong. He served for many years in Hong
Kong's Legislative Council and Executive Council. He contributed to the
Sino-British negotiations on the future of Hong Kong and has served as
an advisor to a series of Hong Kong governors. In the business world, he
is the chairman of P J Lobo and Company and the director of a further
seven companies based in Hong Kong. His many honours include the
Officer of the Order of the British Empire (OBE) and the Commander of
the Order of the British Empire (CBE).

Much of Sir Roger's account focuses on his childhood and his time in
Macao during World War II, when he supported the efforts of Allied
intelligence—an activity about which he has kept silent, until now.

I STILL THINK OF MACAO as a charming, interesting place but a place
that has changed a lot. If you put the clock back to my childhood, more
than seventy years ago, Macao was very much a colonial outpost, even
given the fact that the Portuguese always assimilated when they
colonized a place. Unlike the Dutch, the British, the Spanish, and others
who took a place and ran it the way they wanted to, where everyone
had to follow their rules. For Portugal, I think the main objective was to
turn people into Christians. Portugal went around finding Christians,
the Spanish went around finding gold, and the Dutch went around

finding trade. That's why Macao was given to the Portuguese and not taken by force.

It was a very Christian place. As a boy, I remember playing, riding a bicycle, kicking a ball, or whatever, and at six o'clock the church bells would ring and the children would cross themselves and go home because they knew it was time to go to church. That was a colonial way of living. I'm not saying all the natives did that, but Portugal was a Catholic country and Macao was a Catholic place.

Macao was divided into parishes, just like Portugal. There was St Lawrence where we lived, which was the most elite parish, then St Augustine, the newly Christian parish and St Anthony, with the ruins, where the firemen and police families lived. It was the way Macao was divided, socially as well as practically. Later on, when I played football, people would say: 'What are you doing playing on their team, with the sons of the soldiers and policemen?' And I'd say: 'They're my friends.'

Macao has moved on and today, of course, it's quite different from my time. The place is over-built. Anything Hong Kong could not do, Macao would do, like greyhound racing, casinos, and things like that, but if Hong Kong could do it in the first place, the money would stay over there.

There was the charm of a leisurely day, walking down the Praia Grande under the trees, going for a meal and a bottle of wine, no one pushing and shoving. Today all that has disappeared. It pains me that it's all very commercialized, everyone's out to make money. In the old days people in Hong Kong could say: Let's get away from the rat race and go to Macao for a peaceful time, a lovely evening, a good meal. Not any more. Waiters slap the bill on the table as soon as you've finished because they want the table for the next group of people.

I studied under the Jesuits at the seminary. When I left school, I wanted to go to the United States but when I told my father this, he said: 'How can you go to the States? You don't speak English.' He was right. We learnt Portuguese, French, Latin at school but never English. I was one of the few children to learn Cantonese at school. I forgot most of it afterwards, although now I'm fluent—I couldn't do without it. The emphasis at school was on Portugal. So much so that in the classrooms, there was a world map pinned up on the wall and superimposed on the world map there was a map of Portugal and her colonies. Portugal was only a tiny sliver of Europe but when you started adding Angola, Timor, Mozambique, Cape Verde, and all those places, Portugal could cover the whole of Europe. And at the bottom of the map, written in red capital letters, were the words: 'PORTUGAL IS NOT A SMALL COUNTRY.'

It was a colonial administration. The education was in Portuguese and everything related to the motherland. There were some Chinese pupils at the school and they had their own classes in Chinese. Even in the seminary, there were some of us learning in Portuguese but others, who were ethnically Chinese, learning in Chinese. The Jesuits were always looking for Catholics to become priests and all these Chinese were Christians. But compared with what children get today at school, we were given a very broad education. We'd talk about everything—the human body, or the stars, or whatever. Nowadays they have to decide at twelve or thirteen if they want to do arts or sciences, and once they get on a certain track, there's no turning back. In that regard, my school days—playing with my friends, riding our bicycles across the cobblestones—were a charmed existence.

My childhood was very Portuguese. We'd wash and go to dinner, always a long set table, and we children would take our places at the lower end of the table. If there were guests, they'd be shown in first. The food was continental, with soup, bread, and olive oil. Someone almost always came for lunch—we were used to that. The people from the same parish stuck together like a clan. Even the way of speaking could be a bit different, from one parish to another. Some a little more polite, some a little more coarse.

Our family was very musical and on Saturdays we sometimes held a musical afternoon where friends came with violins or cellos and we had to perform. A hateful thing! We'd be dressed up in sailor suits or something and had to sing a song or recite poetry. It was all very cultural but as a seven-year-old boy, it was pretty horrid.

The Portuguese did try to integrate but there was always a divide between the Chinese and the non-Chinese. I suppose the Chinese didn't want to integrate too much either. Business people mixed at a certain level because they had to do business, but the Chinese were Chinese. There were Chinese districts, Chinese clubs, Chinese restaurants—they had their own things. It was almost like two separate worlds. Of course there were some cross-cultural marriages, and after a while there were more local people marrying Chinese. But the real, real Chinese were the fishmongers, the junk people, the tradesmen, the shop keepers, and the working class in Chinese society. We all knew them, we all met them, but there was very little social contact.

My father came to live in Macao from Portuguese Timor. On my mother's side they came from Scotland four generations ago, settled here and became school teachers. So I'm not one of these ten-generation Macao people but I do consider myself very much a Macao and Hong Kong person just the same. My father started by working in the bank and doing government jobs. He got permission to trade, and he did well. Like many people in Macao, we weren't terribly rich, not like Hong Kong tycoons, but we had a very comfortable, good quality of life. To me, being rich means having as much as you want and being satisfied—and we were. I opted to leave Macao because it was becoming too parochial. I had big ideas about more democracy, giving the people more say, more localization. I think I was born before my time! So I opted for another lifestyle and I never looked back. But those who come from Macao always, through blood connections or whatever, feel a sense of belonging to Macao.

In 1941, when the war came to Asia, I was eighteen. I'd been over in Hong Kong trying to brush up on my English so I could go to America. But then war broke out and I went back to Macao on one of the last ferries out of Hong Kong. In some ways, the war years were pretty good—a lot of girlfriends, a lot of friends from Hong Kong. We played tennis, we went dancing, learnt the jitterbug. Macao really opened up with so many refugees. It was great. People didn't have money in their pockets but the spirit was there and the fun was there too. People were starving but we were not at war; we were just suffering the consequences of war.

I found work at first at the Portuguese aviation centre. They had a naval base with five or six planes and I worked there as a fitter-mechanic. I flew up with them and washed the planes. It was OK—I was just a kid. The planes were propeller driven, small two-seaters with huge floats, the kind of things they use on rivers in Canada. The pilots, four or five of them, were all senior officers from Portugal. The hangar was just where the hydrofoils are now, but the harbour itself was bigger in those days . . . there's been reclamation since then. There used to be a ramp right there in front of the hangar and we pushed the planes on cradles down to the water, making sure they got into the water, then pushed the cradle back. The pilots used to play practical jokes on us. They'd ask us to go and try to start the engine, which meant we had to stay on one of the floats and try to pull the propeller down to start it up. They'd wait till we were down there, then pull hard on the throttle, and leave us standing up to our waists in the mud or flailing in the water and we'd have to swim back.

Later I got the chance to work at Macao Electric. I was always mechanically minded. I worked on a team there, learning how to put together turbines. I did night shifts, making sure Macao had enough electricity, that the power level was up, working alongside the coolies and mechanics. That's when I learned to speak the other kind of Chinese, the kind not used in polite society. Then a friend of mine approached me and said: 'Look, why don't we go into the rice business? But first we need to get a machine to polish rice.' I said: 'Let's look around and see what we can find.' I found an old generator and diesel engine left in the naval dockyard, and thought: maybe we could improvise something. So we did improvise, and soon we were the only people in Macao able to polish rice.

We got rice from China, and we would trade wet rice for the dry rice that we could polish. We learned our trade: when to go for a very fine polish, when to leave a little bit of husk, for the children with beri-beri, for example, and what to do with the broken rice. You could sell anything in those days, even the husks.

I was also involved with British intelligence, through my father. That's one thing we've never discussed when we've talked about our lives. We were part of the British Army Aid Group, sending back into China pilots who were shot down around this area, or who found themselves in Macao. We'd push them back into Kunming to rejoin the airforce. I got quite involved in that. The intelligence group would pass information on to us so we could go along and meet the pilots at an arranged time, make sure the coast was clear, and see them board the junk. We had to send two junks, one with gasoline and other goodies and a noisy engine, designed to be caught by the Japanese, and the other one that the people were hidden on to go right through to Free China. It was risky for us, risky for them, risky for everybody, and we learned not to talk about it. We'd go and play tennis, then at ten or eleven o'clock, I'd go along. Our group was part of Mountbatten's Free China. The Americans used to be shot down and come through the same route because they were Allied forces.

I never felt hatred of the pro-Japanese Chinese or the Japanese in Macao. To me they were just people doing their thing and I was doing my thing. The time I felt a bit of disappointment was when I saw Mr Fukui, the Japanese Consul General, who was a great friend of Mr John Reeves, the British Consul—they lived next door to each other—and when war was declared they shut their windows so they didn't have to talk to each other again. In fact, the Japanese Consul was shot by his own people because he was too friendly with the Allies. I went and asked the doctor: 'Can I go and have a look?' because I had to report back to the British. I went in and there was this man, already dead, bullets all over his stomach, and the doctors digging into him to get the bullets out. The way they were doing it! Cutting bits out and throwing them in the bin and all that. I wasn't only upset about what they were doing to a human body but I was angry that people could just eliminate another person like that. To me, life is a precious thing—live and let live.

When I met with the British Consul, we'd go drinking and he'd take a glass and show me something he'd learned to do in England, to chew up the glass as well. Imagine! I felt completely out of my depth. But I made a lot of friends during those war days. Refugees from Hong Kong, friends who worked alongside me with British intelligence. I met my future wife, too. We had our risky moments, like going out and having a few drinks, boys and girls getting together, and suddenly realizing one of the girls was the girlfriend of the Japanese general. He burst in with his bunch of heavies, they were drunk too, and started waving their swords around. We ended up looking pretty silly.

Probably the most dangerous time for me during the war was when the Americans came and bombed the aeroplane depot in Macao. There was a hangar but it was being used as a store for gasoline and other things we could trade with the Japanese for rice. One morning the Americans came in and they zoomed over Macao, something they'd never done before. They zeroed in on this place and—bang!—they hit it. I was with my father, it was seven o'clock in the morning and I said: 'I'm going down there—because all our stuff's there.' This is what we were using to trade for rice: gasoline, church bells, metal frames, wire, nails, anything we could get our hands on. If you had a big church bell, they'd buy it!

I rushed down to see what was going on and the planes turned round and came back. I had my motorbike right at the door of the hangar and my father zoomed up behind me with his car. They didn't just shoot at the hangar, they shot at the cars, the motorbike, everything. We started running all over the place, we hadn't bargained on that happening. Then the whole thing went up in flames. I saw my father running away— I'd never seen him run before in his life!—but his car was shot out. Amazingly, neither of us was wounded. That same afternoon, the Americans came back again, just to have a look at the damage, and some of the Macao policemen took out their peashooter pistols and tried to fire up in the air at the planes—tak! tak!—absolutely ridiculous.

Those years went quickly for me. Before you knew it, you became old for your age, very serious-minded in many ways. If you were in a prisoner-of-war camp, the war years must have seemed endless. But if you were free to move about and you were a young man growing up, life moved very quickly. Before we knew it, the war was over. As the war was ending, I was sent by British intelligence to Hong Kong on a junk. There were four of us altogether, each with different instructions. It took forever for us to get there and when we finally did arrive, the Japanese were all over the place, but they were surrendering. We saw the Allied troops coming in, the planes coming in, and we went into the camps. Each of us had sets of instructions to pass on to various people, about what they were going to do now. By that time, after all those experiences, I was an old man. I'd lost my youth, lost my carefree attitude to life. I'd turned into an old man at twenty-one.

I feel sad about the handover, not terribly emotionally sad, but sad because Macao has already lost a lot of its character. Macao in the years to come probably won't have strong leadership; no real leader has emerged, so far. Unlike Hong Kong, which prepared for years, Macao hasn't. There's been almost no localization here. What Macao needed most was to have its own people running the place, but that never happened. Very good people, the really talented ones, became doctors or lawyers. No one went into the civil service, because they weren't accepted there; civil servants had to come from Portugal. So the general sense was that the expatriates would come and take all the senior jobs and local people would never get beyond a certain level.

But as for the handover, what have the Portuguese actually given Macao, what have we given the people? Have we given them a sense of security? I don't think so. Have we given a place that's corruption free? I don't think so. Have we given a place with law and order? I don't think so. Have we even given a place where you can go and take your family for a weekend? I wonder. What is there to do in Macao now—go to the casinos, play golf? I think it will take some doing to make Macao continue as Macao. I would hate to see Macao become a small Chinese village because in terms of population, that's all it would be, an inconsequential village. What has Macao got to offer, to make it as viable as Hong Kong? A small airport? A few casinos? I feel very concerned, in some ways sad, that we haven't done more to give Macao continuity.

I don't mean it should carry on having a colonial administration or continue as a Portuguese place but its charm, its traditions should continue. But I think the strength of Chinese culture will erode all that. I've talked about this to a past leader of Macao and asked: 'What exactly are we leaving behind?' and he really disappointed me by saying: 'We're leaving them culture.' I said: 'What culture?' What culture have we got to give that can compare with China's terracotta soldiers?

Unlike Hong Kong, there is no international community established in Macao. So the forces of the Chinese will eventually take it and turn Macao into a place more Chinese. Good luck to them. But it's a pity, too, in my opinion, because, being nostalgic or romantic, it was a charming place, the place I grew up in, a place full of memories. I'm not painting a bleak picture of a place without a future. It's just that it will be a very different future.

Chui Tak-kei

Honorary Citizen

C hui Tak-kei is one of the great success stories of Macao's Chinese community. He was born in Macao in 1912 to a humble family whose ancestral origins can be traced back to Xianghui in Guangdong province. He grew up at a time when opportunities for members of the Chinese community to progress in Macao's society were very limited. Nevertheless, Mr Chui rose to become both a prominent commercial figure and a key political player, an influential voice in the preparations for Macao's return to China.

Mr Chui started his career intending to become a pharmacist—but became, instead, one of Macao's best known civil constructors. He began work as a civil constructor in his twenties. His legacy is visible throughout Macao's architectural landscape. One feat of which he is most proud, was building Macao's first high-rise building, a ten-storey block in central Macao, completed decades ago but still in use today.

Mr Chui has played an important role in the local Chinese community since the 1950s. From 1953 to the present day, he has served as the President of the Tung Sin Tong, a well known Chinese charity organization providing relief to the poor. As a business leader, his influence is also strong. He is the vice-president of the Chamber of Commerce in Macau, a position he's held for many years—and has served as the President of the Civil Construction Association from the 1980s to the present day.

His mainstream political career developed in later life. He was vice-president of the Legislative Assembly from 1976 to the early 1980s, and has been a key figure in the negotiations on Macao's return to China. He was the president of the Consultative Committee of the Basic Law, Macao's post-handover constitution, and a member of the Basic Law drafting committee. He was also a member of the Preparatory Committee, which laid the groundwork for the selection of the Chief Executive.

Chui Tak-kei has been awarded a number of honours by the Portuguese government for his services to Macao. He was awarded an honorary PhD from the University of East Asia, now the University of Macau, in 1986 and became one of only two honorary citizens of Macao in 1998. His memories of Macao's war years and his own account of key political events, such as the 1 2 3 incident, make a sharp contrast with those of the local Macanese community.

I CAME FROM A POOR FAMILY, nothing fancy, nothing rich, very humble. I had to work to survive. The schools I went to were run by the local charity committee and offered free tuition. My family couldn't afford to pay for me to study. I only studied English for a few years so my English isn't very good, but I studied Portuguese and of course Chinese for a long time—and learned a little French, too.

Our home was very simple. A small flat on one floor, no stairs. Very humble. It was a traditional Chinese home. My father was also in the construction business and influenced me a lot. I remember from an early age watching him building houses. He was a mason—and it was hard work. My father made about 90 cents a day—that was all. I was just a child then, and now I'm eighty-eight, so you can imagine how different life was.

Macao was a town with no electricity, only gas lamps. The roads were rugged mud streets. There were hardly any concrete roads like we see today, scarcely even one like that. The first concrete road was the one right through the middle of town, the main artery that runs alongside the Leal Senado. Later they started using cobbles, following the road design in Portugal but made by the local Chinese population. Soon they were nearly all cobbled roads. You can still see some of them, in the smaller streets around the ruins of St Paul's Church.

We lived in an all-Chinese district, a small community where everyone knew everyone else. It was almost like a village. It was a nice, peaceful neighbourhood with decent people. I came from a happy home. Our family life was simple, and we were easily satisfied with what we had. There wasn't much to do then, not like nowadays, so we didn't aspire to any more than we already had.

I went to Guangzhou to study civil construction. At that time my family was doing a little better financially so they could afford for me to go away to study. My first love had been pharmacy. From seventeen to nineteen I had a job as an apprentice in a pharmacist's shop in Macao, and from nineteen to twenty-one I went to pharmaceutical college in China. I really loved pharmacy and wanted to make it my career, but the next step would have involved going to Shanghai to study more advanced courses and we just didn't have the money for that. So I switched to vocational school and studied construction. I was influenced by my father and my uncles who all worked in the construction business. I didn't have much choice really. They pointed out that even if I managed to qualify as a pharmacist, I would need a lot of money to open a clinic. So it made more sense to take their advice and choose a more practical career.

Even so, it wasn't easy to survive. There wasn't much work in those days in the construction business. Luckily for me, my uncle ran a restaurant so when building work dried up, I could always go along there and help out. I was a hard worker. At that time I was also taking morning classes in English and going to night school in the evenings to study.

The years of World War II were tough. We weren't really involved in the war in Macao, but it still made daily life in the city difficult. The food supply was low and medical supplies were in scarce supply. The Tung Sin Tong charity organization set up special kitchens and handed out free rice soup to people. They also set up clinics, offering Chinese medicine. I started helping out there. They were hard times. Hundreds of people died from hunger. I don't remember going hungry myself because our family had the restaurant. The authorities discounted the price of rice supplied to the restaurants to make sure the restaurants stayed in business and the food supply was maintained. The Japanese invaded China and started the war. Numerous innocent citizens were killed.

The 1 2 3 incident in 1966 took place at the same time as the Chinese Cultural Revolution. When it started, there were many people in Macao who supported the Kuomintang. Rioters started spreading rumours about the Portuguese not treating the Chinese properly and things got tense. The Portuguese had their army stationed in Macao then. They declared a curfew across the city after the incident. The conclusion of the crisis was that the Portuguese signed an agreement with China and the whole thing came to an end.

I was involved with the Chinese Chamber of Commerce at the time. Mr Ho Yin was the Chamber's president and went to negotiate with the Portuguese government on our behalf. I was one of his group of advisors in the Chamber. There were two Portuguese leaders representing the government side. The Chinese government wanted the Portuguese to apologize to the people of Macao and to give proper compensation to the families of the people who were killed. The conclusion of the negotiation was that no more KMT flags would be raised in Macao, and all KMT organizations would stop operations in Macao. It was a difficult time. Everyone was hoping for peace. Things did improve for the Chinese in Macao as a result of the confrontation but even afterwards it wasn't as open as it is today. The administration was more like a dictatorship.

Later, in 1974, the Portuguese had a revolution themselves and the system did relax after that. Before the Portuguese revolution, they used to describe Macao, Taipa, and Coloane—the two islands that form the territory of Macao, along with the land at the tip of the peninsula—as three provinces of Portugal. After the revolution, they changed the terms and called them Portuguese territories. They treated Macao like a colony. It was a dictatorial government. Only gradually, as more Portuguese came to Macao, did they start to get along with the Chinese community better, and the relationship improved. In my eighty years of life, I've seen a slow warming of relations between the Portuguese and the Chinese here. They gradually treated us better. Few Macao people studied Portuguese, but most Hong Kong people studied English.

The return to China at the end of the year makes perfect sense. The important thing is to make sure the one country, two systems formula is implemented properly, so that there is no change for fifty years. If that is respected, I can't see any problems. I don't know what I'll be doing on the day of the handover because I can't be sure of my health nowadays, but if we do finally have a system of Macao people governing Macao, of course I'll be very happy to see it.

Jose Chui

East Meets West

J ose Chui is a son of Chui Tak-kei, the powerful figure in the Chinese community whose memories are recorded earlier. Like his father, Jose Chui, now in his late thirties, has become an influential civil engineer in Macao, and is tipped to be a leading figure in the future.

Although the family's ties to their ancestral village of Xianghui in Guangdong province remain strong, Jose Chui is now the sixth generation of his family to be born in Macao. Much of his childhood was traditionally Chinese, but the great success of his father, combined, perhaps, with the gradual breaking down of the old divisions between the Chinese and Macanese communities, gave Jose Chui many more social and educational advantages than his father had enjoyed.

Mr Chui studied and worked for some time in the United States. This exposure has helped him to develop an international perspective alongside the strong sense he has of his Chinese identity and cultural heritage. He is also the first person in his family to convert to Christianity, in contrast with his family's traditional Buddhist practice.

OUR FAMILY ORIGINALLY came from Chui Ka Hung, Xianghui in the south part of Guangdong province. It was a very poor village of about two hundred people, all with the same last name, Chui. There was no chance of building a better life there. So one day our great-great-grandfather thought: 'Let's take a chance and see what happens.' And that's how we came to Macao.

Even now, every Chinese New Year, we go back to our home village. It's a big event. The village is very small, still only about two hundred people. They have paved the roads and brought in electricity and water, but apart from that it hasn't changed. It is hard for me to relate to the place because none of us grew up there. It is like looking at a time tunnel—this is what you would have been if someone else hadn't made a decision for you a long time ago. The villagers enjoy it when we visit. The kids come running out and we give them candies and crackers and all that. The old ladies and kids greet us because the men are all working in factories now, they don't work in the fields any more.

Although I was born in Macao and grew up here, there's no doubt about it: I'm Chinese. I accept the fact gladly. Macao is my home town, but even though we've been away from Xianghui for so many generations, I'd still say that's where I'm from. Although deep down I think of Macao as my home, I don't make any special effort to put my Macao roots and my Chinese identity together.

My childhood was very quiet. My family had a large house. My sister, brother, and I would play inside the house; my parents wouldn't let us go out to play in the streets, even though in those days the roads were much quieter. Even when we rode our bikes, we had to do it inside the house. It was big enough; with a full-sized bike we could make turns and everything. We had a lot of servants. It was a three-storey house, with a garden slightly bigger than the house, with swings and slides, and lots of different kinds of trees: mango, and orange, and guava, and some quite rare Chinese fruit trees. And fish ponds, too. It was a very comfortable set-up. It was a good neighbourhood. There was a wealthy Portuguese family living all around us—they had four or five houses there for their sons. There was also a monastery next to us. It was very quiet and safe. In that sense it was a slightly mixed neighbourhood, which was unusual.

My parents used to invite many friends home. My father loves painting, and in the evenings he'd invite a large group to the house and they'd paint until quite late. My mother used to like singing Chinese opera, so sometimes she'd invite a Chinese band to come and play in the house and people would take it in turns to sing. My sister and I would walk round listening to what was going on. Guests would stay until quite late and we wouldn't have to go to bed until about ten o'clock in the evening.

My father never had a traditional Chinese breakfast—it was always a cup of tea and some toast. He's been working with the Portuguese for a long time and speaks very good Portuguese. In some ways, he's quite liberal but in others, he's quite traditionally Chinese. When I was in my twenties and living in the United States, he wanted me to come back to

Macao to work with him. I remember him saying: 'Everyone else has his son close—it's just my son who isn't.' But he never forced me to come back. When it came to marriages, having children, my career, he let me make my own decisions. That was quite liberal.

When I was a boy, there were only two or three schools for the Portuguese children using the Portuguese language. All the other schools taught in Chinese. When I started to learn a second language, it was English. It was unheard of to learn Portuguese. The only reason we went near a Portuguese school was to use its football pitches, which our school didn't have. After school, we'd go up there. The priest wouldn't let us in so we had to hang around until he left and then climb over the wall to play there.

The Chinese and Macanese didn't interact much, although my father had many Macanese friends, so I never had bad feelings towards them. They were brought up in their culture and we were brought up in ours. It's a shame we didn't mix well together. I never felt that the Portuguese shouldn't be in Macao. We had a different lifestyle, and different traditions, but we both belonged to Macao.

I left home when I was fifteen. It was a beautiful misunderstanding between my father and me. Somehow he got the impression I wanted to go overseas and study, but I thought it was what he wanted, so I agreed. I was sent off to high school in Hawaii. The first year was really tough. I'd only ever been to a Chinese school and my English was pretty bad. I had to look up every other word in the dictionary. I remember my first cousin taking me to an ice-cream parlour and the only thing I could order was: Black and White. They were the only words I could recognize and the only words I felt confident about pronouncing.

Looking back, I think it was a very good move. The high school experience was important in building my basic communication skills. If I hadn't gone to an American high school, I'm not sure I would have been accepted into an American university, the University of California, Berkeley, otherwise. I completed my graduate studies in 1983 and then worked in San Francisco for some time, including helping to rebuild the city after the earthquake in 1989. I came back to Macao in 1990, mostly because my father really wanted me to.

When I returned, I found my perspective had changed. For one thing, my religious beliefs were very different. I had become a Protestant. My family are all Buddhists. When I went to Seattle, I got involved in Chinese Fellowship—I hesitate to call it a miracle but before that, I was quite sceptical about Christianity. I went to a Chinese Christian school in Macao but it didn't convince me at all, but in the States I was debating with the pastor one day and I heard a voice saying: 'Joe, your logic doesn't work here.' And I knew it was right. Faith isn't logical, it's like love. At first my family didn't like the fact I'd converted but now they have accepted it. We don't make a big deal out of it.

I was also made aware that the way of thinking in Macao is different from that in the United States. Here people want to get to know you personally to do business. It takes them a long time to feel comfortable with new people. My father is in the construction business, and he does some development himself, so often, when people know who I am and know my father, they would refuse to give me jobs. They seemed to think: if you get involved in the design of this building, you might end up taking over the whole project yourself. It wasn't a big plus having such a famous father.

Also I got used to being quite casual in the United States, but here people had definite expectations about how I ought to look and behave. They would say: 'You ought to dress properly in a suit and tie.' In the States, blue jeans and a T-shirt would be fine for an engineer, even for the boss. Even now I don't always comply. If I go along to a site, I'll dress casually. It's hard to explain—you can feel under pressure. Even saying: 'I'll do what I want to do, I'm my own person' is completely foreign to the thinking here.

On certain other topics, I'm still very traditionally Chinese. I want my wife to stay at home, not go out to work, for example, while the children are little. That makes me sound very traditional. A lot of Chinese people think rich people should hire servants—chauffeurs to take the kids to school, tutors to teach them after school, maids to wash them and feed them and put them to bed. But I think you ought to do everything yourself. Would you give your wallet to someone else? Of course not. But surely your children are worth more than your wallet— so why would you hand them over to strangers? Their best teacher is probably their mother. With my wife looking after the children, I expect them to grow up as friends, and teenage problems can be sorted out before adolescence. Teenage rebellion only happens if you can't communicate with your parents. We only have one servant who cooks and cleans so my wife can spend more time with the kids. Sometimes when local people see my wife go to the market—and occasionally I go along too—they're really shocked. In that sense, we're quite different from a traditional Macao family. A lot of wives take pride in the fact they never have to touch their kids. They think that shows their class. I don't agree with that.

My children's upbringing is very different from my own. My father worked long hours and also served in the Municipal Council, so he was busy, and in the evenings he socialized. So we didn't communicate much. I remember when I was about four years old, he bought ten volumes of Chinese stories and read every one of them to me. It helped me to learn more about Chinese culture—and also brought us closer. Now I read my sons the same books.

My wife and I met when we were about ten years old, at primary school. When I went to the United States, we kept in touch and wrote often—and when I went to Berkeley, she came and joined me and went to college, too. We graduated in 1983 and got married the day after my graduation. My father was getting old and I felt he shouldn't make too many trips to and from the United States, so I thought: while he's here for my graduation, we might as well get married. So in the space of two days, I became a graduate—and a husband!

The handover transition doesn't worry me at all. Macao is always in the middle of the road. Even at this great moment of returning to China, Hong Kong has gone ahead of us and Taiwan is still waiting, so once again we're in the middle. I don't expect anything dramatic to happen. The basic system won't be changed for at least fifty years and we'll still be governed from Macao. Fifty years is almost like an eternity. In Chinese, there's a saying: 'It takes ten years for the world to change completely.' So fifty years is ages. I don't think we have to worry. In fifty years' time, China may be more liberal than Macao. If we close our door for fifty years, it may be us trying to catch up with China, and not the other way round.

My theory is that once people have seen something better, there is no way they will turn back. In China, many major cities and provinces have seen something better. Unless blood is shed, people won't turn back. China is going to change in a way that will bring it very close to the Western world and Macao's going to hitch a ride and go along with that. So why worry?

In principle, I think the Portuguese and the Chinese are similar. They enjoy personal relationships, they both like to have theories and talk a lot, and they value tradition greatly. Efficiency may not be the highest priority. So actually we didn't have to change much to adapt to the Portuguese way of life. That's why there are so many intermarriages in Macao. On localization, I think we have to be realistic. When it comes to power, no one wants to give it up. To a certain extent, you can't really blame them. Before 1995, a quick localization process definitely would have helped, because people would have been in their posts for at least four years by the time of the handover.

But now, with so little time left, I'm not sure it's such a wise move. There's no point emphasizing quantity over quality. A smooth transition might not be helped by a change in personnel. In Hong Kong, the emphasis on localization is at the top posts—even now, you still have sergeants who are British—but in Macao, the process is quite different. We started from the bottom, replacing the lowest-ranking people, then the next lowest-ranking people, and so on. If you start from the top, it's more effective and much simpler because you are picking out the best and the brightest for certain key posts.

Something has to be sorted out between the Portuguese and the Chinese. Of course it's the job of the Portuguese, but the people they're governing are Chinese and the Chinese may be more effective at solving problems. If a Portuguese police officer tries to mingle with the Chinese, they know he doesn't belong and the language barrier might hinder him, too, so he has to rely on Chinese detectives. The Chinese government should definitely get involved in some way, before the handover. It's nothing against the Portuguese—but a partnership could make the transition better.

Most people here are more worried about law and order than about the handover. The situation is pretty serious; it affects Macao's strongest sector, tourism. It's all very well for us to say that this is just gang fighting. To an outsider choosing where to go on holiday, it's a negative factor. I might go to South Africa now, but I wouldn't have gone ten years ago. You only have so many days in a holiday and you want to choose somewhere safe. The world economy has been a major factor—

the general economy has been worsening since 1993, so people who are surviving on casino earnings are running out of money. They have to find ways to keep the income level up so they start bothering normal businessmen and that gets very serious. There are certain things China can do that people will fear, but as outsiders, they can also act on a lot of things that local police are afraid of. Until we can improve law and order the public image is going to be very bad. Every year we spend millions of dollars promoting Macao but you only have to see one evening news report showing a murder in Macao and that wipes out the million dollars.

My kids are going to the same school I went to. I'll probably send them to the United States for high school, taking the same route I did. There are still no Portuguese in their school, and no Macanese, just one or two kids with fathers who are English or American but mothers who are Chinese, so they just don't mix with non-Chinese either.

I think my kids will have more opportunities because of the handover. They can go all over China and, of course, being from Macao they'll have advantages too. Many company laws are still not very clear—if you're an investor from Macao in the Mainland, for example, are you treated as a foreign investor or as a Chinese? What rights do you have? That remains to be explored.

On the day of the handover I think I'll feel quite emotional. For the first time, I'll be legally Chinese and I think when I hear the national anthem played here, I'll find it quite moving. It's part of history that China will take back all of its land—it is a natural course of events. There's nothing good or bad about it, it's just the clock ticking.

Henrique de Senna Fernandes
A Writer Remembers

Henrique de Senna Fernandes, a well known lawyer is also a successful novelist. Such novels as 'The Bewitching Braid,' recently made into a film, explore love affairs between Portuguese men and Chinese women in Macao, especially in the pre-war period.

His family has a long and distinguished history in the enclave but had fallen on hard times by the time World War II broke out. As a result, his personal memories of the war period describe hunger and suffering, set in a rather manic atmosphere of wild parties, fuelled by the sense of uncertainty. They make a stark contrast with other wartime accounts in the section, a reminder that although Macao is a small place, the differences within its society can be considerable.

His lyrical sense of regret for the changes in Macao, both physical and social, make it clear why he has been such a success as a writer.

WHEN I WAS BORN HERE, many years ago, Macao was a quiet town, sleepy but very beautiful. It was divided into two parts, two different cities: the Christian city, where the Portuguese community lived, and the Chinese city. You could see a clear difference between the two, just walking around.

The Chinese city was where business was done. It was very compact; if you walked through the streets there you saw houses pressed in tightly against each other, but no trees. Macao's commercial life was entirely in Chinese hands. We went to the Chinese part of the city very often, to visit and to go shopping. The Christian city was very small in comparison. We were maybe 10,000 Christians and the Chinese population was tens of thousands. The way of life was different. The Christian city was much quieter, with traditional cobbled streets, private houses which always seemed to be having parties, secluded avenues, and gardens. It was divided up into districts, each one of which had grown up around a different church—St Lazarus, St Anthony, St Lawrence. . . . Everything was named after saints.

It was peculiar to sense where these quarters began and ended, each with such different types of people, with different ways of life. Such a small city, but one moment you're in a residential district, the next you cross the street and you're in a commercial area, then cross another street and it's quiet again. Those are the contrasts of Macao. It was very beautiful and I loved it.

My family has been in Macao for about 250 years and I have some Chinese and Portuguese parentage. My ancestors came from Portugal originally. I remember my mother telling me about a Frenchman who was perhaps a great-great-grandfather, I don't know exactly how far back he was in the family tree. We've lost all trace of our ancestors. In Macao, the archives are only compiled by the church and many records have been lost forever because people in those days didn't care for them. My Portuguese ancestry was an important part of my education. Our education system was completely focused on loving the motherland, Portugal. We were taught to love a country we'd never even seen. The first time I saw Portugal was after the war when I was twenty-three years old. It was very different from Macao, and in my dreams I'd imagined it quite differently.

The education system was quite wrong in putting so much emphasis on the motherland and completely ignoring life here. It was sometimes so stupid. We were forced to learn by heart the names of all the railway stations in Portugal—not just the stations but the substations in the tiniest Portuguese villages. We're talking about children who'd never seen a railway, let alone a train! It was compulsory. And we had to learn the names of all the small rivers in Portugal—but there was no mention of China's great rivers. That was all wrong.

I didn't learn Chinese at all at school. Unfortunately, although there was one Chinese class, I wasn't interested and no one encouraged me to learn Chinese. It's a tremendous frustration for me now not to be able to read Chinese. I see the characters all around me and I can't understand them. It's a pity because I love reading, I read for two hours every day. If I could read Chinese, I'd have a better understanding of Chinese thought, but it's too late now. I know a few characters, of course, but nowadays I find it so hard to remember new ones. I don't blame my parents for not encouraging me to learn Chinese—it was the thinking in the Macanese community at the time.

I was born into a wealthy family, but life is very peculiar; my father lost all the money and we became very poor. I know exactly what it means to be poor. My father had his ups and downs, as everyone does, but we survived. The hardest years were the war years. The war began in 1939 in Europe, but for us, in this part of Asia, the war began in 1941 when Japan attacked Pearl Harbour and occupied Hong Kong, Singapore, the Philippines, and the Malay peninsula. The Japanese never occupied Macao, but they were all around us, they were a reality.

They were dreadful times. Macao is a tiny territory, and we were overflowing with refugees. The population swelled to about half a million people, mostly refugees from China, many of them practically beggars. We opened our doors to anyone and everyone. It was characteristic of Macao—an open-door city. Nowadays it's very different.

I was eighteen in 1941, my final year of high school. I had planned to go to university afterwards but suddenly the war came and I couldn't. I ended up losing five years, five years without studying at all. During the war, I had to work so I could help the family. We really needed the money. I found work as a teacher in primary school and I also worked for the post office.

We witnessed suffering we couldn't even imagine. Misery. Brutality. The realities of war. The hunger. It was like the worst kind of nightmare. I saw many, many beggars here who looked like skeletons. Nowadays it's not unusual to see famine victims, starving African children. I saw that during the war. At the end of the war, they showed news footage of the horrible scenes at Belsen and Auschwitz. Those poor people, so thin and so maltreated. I wasn't very surprised or horrified because I'd already seen it—in Macao.

I was hungry, too, during the war. We would maybe get a bowl of rice, occasionally with some Chinese shrimp paste on the top, or congee [rice porridge]. It was never enough to live on. The worst years were 1942 and 1943. By 1944, it was better because you got used to the war, you didn't feel so sad or so desperate, because you were used to it. Also, by 1944 I'd started teaching and I was earning more so we managed to live better.

The war years in Macao were terrible, but it's strange, I've never had such happy times either. Every night there were parties. Parties with no food, no smart clothes to wear, just dancing and having fun. Things were so uncertain. We wanted to forget all the gloom because we never knew what tomorrow might bring. Macao had never known such wonderful orchestras, which had come across from Hong Kong from the cabarets and night-clubs there. We never stopped dancing. Every chance we got, we thought of an excuse to have fun, to dance.

There was always the threat that Japan would invade and occupy Macao. In 1942 and 1943, we didn't know who would win the war, but by 1944 we started to think the Allies would. Even so, the Japanese were unpredictable and you couldn't trust them. In 1939, I lived in Canton [Guangzhou] for a while, in the British concession of Shamian, doing business for my father. Every day I used to go to the Chinese quarter, which was occupied by the Japanese. They invaded South China in 1938 and occupied Canton in the autumn of 1938. I saw first hand the brutality of the Japanese. I saw Japanese guards beating an old lady with a stick, like complete animals. She had tiny bound feet so she couldn't walk very quickly and they beat her. Brutal.

The Chinese community faced difficult times, too, but we had a very good police force during the war and the whole city united, Chinese and Portuguese, against our common enemy—the Japanese, and the collaborators. Some people grumbled occasionally and said things could be done better but we had good police who sought out the collaborators. They were the worst people, the Chinese who collaborated with the Japanese, even worse than the Japanese.

I had to wait until 1946 for a chance to carry on studying, when I finally went to Portugal. In my first year there, I couldn't get used to it at all, I was very homesick for Macao. I didn't like Portugal at all. I had gone there to study and I stayed for eight years and by the second or third year, I did start to love Portugal for itself. Five years is a long time for a young man to wait. I found I had forgotten how to study, lost my touch. I studied law, but I found it very difficult. I discovered too late that my vocation wasn't law but teaching. I don't regret my career in law—it's a profession that gave me good earnings and an independent life, but I was never a brilliant lawyer because my real vocation was to teach and write.

In my novels, I write about the personal relationships people had in the 1920s and 1930s. To me, the 1930s were the heyday of the Macanese way of life. Afterwards, the war came and the Macanese began to migrate—to Portugal, Brazil, Australia, the United States, and Canada. Entire families disappeared from Macao. But in the 1920s and 1930s, the young people who wanted to go abroad, went to study in Portugal, and then came back. The war changed everything, ended our peaceful, quiet life and the whole patriarchal structure. Families used to be complete—grandparents, parents, children, grandchildren, and so on. They all lived here, this constituted their world.

After the war, people lost their sense of security in Asia and then the migration began. It was a haemorrhaging, lifeblood flowing out of a body, it never ceased. By the 1950s, Hong Kong had developed so much that all the Macanese youth wanted to go there, and suddenly we lost hundreds of boys and girls who could have stayed in Macao and married and had children here. Instead they migrated to Hong Kong and to other parts of the world and never came back. As well as being a lawyer, I was also teaching and in those years I lost many, many students. I used to get furious because some of them were so intelligent and might have built a great future here, but what could you do? They went abroad. That went on and on, right through the 1960s and 1970s too.

The enchantment of this city is that everyone knows each other, we're all friends. I can get a new passport in less than twenty-four hours; where can you find that anywhere else in the world? Everything is convenient. From where I'm sitting, the law courts are 300 metres away, the hospital is round the corner, the post office is right outside, and so on; hotels, restaurants, you can almost reach out and touch them. We have everything we need, and anything we don't have, we can get in Hong Kong, just an hour away.

It's hard to explain in a few words what it means to be Macanese. We have our own way of life here. We have memories and traditions lasting hundreds of years, even our own cuisine. We are part of the city and we always will be. You need to spend time in Macao to understand why we are all in love with this city. I could go to Portugal any time, I've got plenty of resources. I've got two houses in Portugal, but I stay here because I was born here, all my ancestors and my children are linked to this city. It's a very strong attachment and it would be terribly difficult to put all that aside and begin a new life.

Sometimes I envy people from other parts of the world because the place of their birth will never see a change of flag, a change of sovereignty. In Macao, that change comes after 1999. It's very poignant to live your whole life believing that the flag, the sovereignty will never change. To consider this place a part of Portugal, and then suddenly to realize that the sovereignty will be changed and Macao will become a pure Chinese city is really sad.

It won't happen immediately. You can't erase four hundred years of history overnight. The Portuguese atmosphere will stay for some time, but nobody can say what will happen in the next century. Hong Kong's handover was such a celebration, all those lights and fireworks, everybody talked about it with optimism. Then suddenly, Hong Kong hit difficult times, and is no longer the same. There is something intangibly different about it. Of course they still have their big hotels and big businesses, but the city has changed. It will be the same for Macao. The Portuguese atmosphere will remain for some time , but change will come. I want to stay to see the handover, and the changes it brings. I feel I can't leave the city I love at such a historic time. I was invited to Beijing by the New China News Agency to witness the signing of the Joint Declaration between Portugal and China. I was there in the Great Hall of the People and I saw it happen. I felt sad, of course, because I knew even then that everything would change.

Our ways are different from the British. The British never cared about making their people belong to Hong Kong, about feeling Hong Kong was really theirs. Even people who were born in Hong Kong call themselves expatriates. That's something I can't understand. Expatriates? Why? How could they not see Hong Kong as their birthplace? The Portuguese are different. We were never obsessed with racial differences, with skin colour. Look at Brazil—a complete mixture of Portuguese, Brazilian Indian, and African. When the Portuguese sailors went overseas, they never took white women with them. They thought Portuguese women should stay in Europe. So they came alone and mixed with the local women, intermarried, and had families. Portuguese men were never racists. There might have been exceptions, obviously, people who were stuck up and remembered their white heritage, but most of them went overseas and didn't care about all that.

That's a point in our favour. We have many faults, we can be disorganized, cruel sometimes, or uncaring, but we are considerate to women, we make love with them, marry them, and have children. That's why you can see traces of Portuguese blood all over Asia. Here in Macao, the Portuguese mixed with the local Chinese from the very beginning. I don't mean with the wealthy middle-class families; the sailors married the Chinese peasant women, women from the fishing boats, and even slaves.

This preoccupation with law and order in Macao is something blown out of all proportion by the newspapers. I really don't understand why Hong Kong is so hostile to Macao. Everything bad that happens here, the smallest things, they always put on the front page. It is a mistake to try to compare Hong Kong and Macao. Hong Kong is a big place, Macao is small. Hong Kong has a big airport, a big stock exchange, big hotels, and a big harbour. Our airport is small and we don't have a harbour at all. What have Hong Kong people got against us? It's unfair. Crime happens here, just as it happens everywhere else. Yes, we have murders, but so does every other city in the world.

What is terrible is the triads, but they're only connected to the casinos. The triads don't affect my daily life or that of the average man in the street. If you meddle in their world, they'll kill you, but if you don't interfere with them, they'll leave you alone. I still go out at all hours of the night because I don't have anything to do with them. I'm opposed to them in the sense that I'm in favour of order and peace in Macao, but I think the reporting is exaggerated. I'm certainly not living in fear as some newspapers imply; we're not all living in a state of terror in Macao. Some parts of the city are bad, but you don't go there. Here, in the main streets, you can walk around long past midnight without anyone troubling you.

As for my memories of Macao, what do I remember? I remember my youth, my days at primary school and high school, my friends from those times, the picnics, my first love, the traditions at Christmas and Easter that are now disappearing forever. I remember my father showing me how to tell if a typhoon was coming by the colour of the sea, the waves, the shape of Lantau Island in the distance. You had to look out for the junks and other boats coming in from the sea. You read the light, the mood of the sky, the clouds—and you knew if a typhoon was forming. Practical things like that were so enchanting.

It makes me sad to remember all those happy times. My father, my mother, the family discussions late at night after dinner when the adults gathered together and talked. Even as teenagers, we weren't allowed to join in with the after-dinner talk. I remember my grandmother sitting there, talking about the old Macao of her lifetime and her friends, long gone. I remember the sound of mah-jong, the cigars, the smell of the drawing room, so many things from the 1930s. We lived in big houses then, every family had its own house. Today one apartment is heaped on top of another, and people don't even know their own neighbours— one building is a small city.

All those years ago, when I was a boy, we could hear the church bells ringing every evening at six o'clock or nine o'clock, the witching hour. Nevermore. We could see the doves flying across the city. We could hear the cockerel announcing the break of day at six in the morning and other cockerels, far away, answering. There are no cockerels now because there are no gardens. And the bats flying low in the sky at eight or nine o'clock in the evening. Even the dragonflies, even they've gone now. And the song of the birds.

We've lost one of the most beautiful things Macao ever had—the white clouds. You can't see the clouds nowadays because the buildings are too tall. Because of air conditioning, you have to close all the windows, but when I was a boy, we didn't have air conditioning, the houses were built to allow fresh air to circulate; it was a completely different way of life. You could see the white clouds coming from the south, and crossing to the north, rolling over us like white cotton wool. The clouds in Macao were so beautiful, and the sunsets, the autumn sunsets were spectacular. We've even lost the sea. We used to have the bay right here, the Praia Grande, now it's all being reclaimed and the sea is far away from us all, we can't even see it any more. We've lost so very much.

Casinos and Crime

Gambling is big business in Macao—in fact, it is the main business. Tourism earnings overall, most of it drawn from the casinos, contribute more than 40 per cent of Macao's total gross domestic product, and the tourism industry, including restaurants and hotels, employs a third of the workforce.

In 1996, more than eight million tourists came to Macao—more than sixteen times its population. Gambling was the main attraction. Since then, the figures have declined, partly because of the economic crisis in Asia, which has caused a general slowdown in regional tourism, and also because of the bad publicity caused by a spate of violent crime in the past two years. This violence is thought to be the result of a turf war being fought between rival triad gangs in Macao, vying for control of the illegal activities—including prostitution, loan sharking, and drug trafficking—that lurk at the fringes of the legitimate gambling industry. Most of the visitors come from Hong Kong, where gambling is strictly limited. Macao gives Hong Kong people the chance to go wild at the gaming table—only an hour away from home by jetfoil or high-speed ferry.

Macao has offered gambling to visitors and locals for centuries, but the nature of the business has changed dramatically in the past forty years. In 1962, the government awarded a monopoly licence for the control and development of the casinos. It was won by Sociedade de Turismo e Diversoes de Macau (Macau Tourism and Amusement Company), usually known by its initials, STDM. Its head is Dr Stanley Ho, the king of Macao's casinos. He describes Macao as having had only two casinos before his company took control of the industry, one in a hotel and one in a private house, next to an opium den. According to him, both were ramshackle. The croupiers were poorly dressed, the atmosphere was decadent, and only traditional Chinese gambling games were available.

Dr Ho soon changed all that. Macao now has nine casinos, and offers a broad range of gambling, including the standard international attractions of slot machines, blackjack, baccarat, and roulette. Traditional Chinese gambling still tends to dominate. Fan-tan is a popular Chinese game, in which the croupier plunges a cup into a pile of buttons and gamblers bet on how many buttons will be left over when as many sets of four have been removed as possible. Another

Chinese game mentioned in this section is Dai Siu ('Big-Small'). This involves three dice, shaken together; gamblers bet on the outcome of the throw.

Macao's return to China isn't expected to change the gambling industry very much, at least for some time. The post-handover constitution, the Basic Law, stipulates that gambling can carry on for the next fifty years, despite the fact that gambling is seen as a social evil and banned in mainland China itself. The more uncertain question is how it will be run. The monopoly licence now held by STDM is assured until it expires in the year 2001. It has yet to be decided whether the monopoly will be preserved after that date.

Some want the casinos to be the focus of more investment in the future and to add more glittering attractions, in the manner of Las Vegas. David Chow, a legislator and businessman with interests associated with the gambling industry, gives his vision of a possible shape of the industry's future. Others want Macao to develop in other directions and to be less dependent on gambling for government revenues and employment.

The casinos and the gambling industry have always attracted opposition from those who see them as immoral and socially damaging. Some see it as ironic that a place famous for being an international gambling centre is also one of the longest established and most passionate centres for Christianity in Asia. The greatest concern about the impact of the gambling industry involves the wave of crime that has hit the enclave in the past two years, badly damaging Macao's international image and causing friction between the Portuguese administration and the Chinese government in Beijing in the run-up to the handover.

Officials in Macao are keen to point out that the crime is amongst organized gangs and has caused no casualties involving tourists or law-abiding local people, apart from a handful of local journalists trapped in one incident in 1998. They also make the point that the murder rate per capita is low compared with many other places. Despite their protestations, the fact remains that bomb attacks, gangster shootings, arson attacks, and point-blank murders of government officials constitute exactly the sort of news that makes headlines—and alarms would-be visitors. The general consensus is that the violence involves rival triad gangs jockeying for position prior to the return to China, possibly with an eye to the expiry of the present monopoly licence in 2001.

Many local people blame the Portuguese administration for failing to keep the peace, and the Chinese leadership has also highlighted law and order as a key current problem. Despite the overall positive atmosphere in the negotiations between Portugal and China on Macao's return, many analysts suggest the question of law and order has clearly become a political thorn. The situation has been complicated by the Chinese announcement that they intend to station a garrison in Macao after the handover. While no one has claimed officially that this is a direct response to the problem of law and order, it has been interpreted as such by many people. Some say the presence of Chinese troops could be a positive influence, acting as a deterrent to the warring gangsters. Others point to the constitutional anxieties caused by the presence of soldiers. China only has the authority, under the formula of 'one country, two systems', to intervene in post-handover Macao on issues of national defence. Macao's internal security should be entirely a matter for the first chief executive and the Legislative Assembly, using the police forces at their disposal.

There is also concern that, unlike Hong Kong, Macao's post-handover constitution, the Basic Law, contains no provision for the stationing of Chinese troops in Macao because none were expected when the Basic Law was drafted. This also meant the practical details of how many troops there would be, and where to garrison them, had to be worked out in the final months before the handover. The Portuguese withdrew their own garrison from Macao in the 1970s and, since then, have relied on the three police forces—public security police, marine and fiscal police, and judicial police—to maintain law and order.

Dr Stanley Ho

King of the Casinos

Dr Stanley Ho is one of the most colourful, controversial, and best-known figures associated with Macao. He's been the de facto founder of the enclave's powerful gambling industry since 1962, when he led a successful bid for the industry's monopoly licence. His company Sociedade de Turismo e Diversoes de Macau (STDM) or the Macau Tourism and Amusement Company, has operated all of Macao's casinos ever since.

The revenue from a total of nine casinos was some US$ 2.25 billion in 1997—a bumper year for tourism in Macao. Given that Macao's government takes more than 30 per cent of gross casino earnings in tax—and that casino taxes make up almost half of the government's total revenues—Dr Ho exerts a tremendous influence on Macao's overall economy. His portfolio also covers shipping, television, the aviation industry, hotels, and a substantial share of Macao's real estate, as well as contributions to such fields as social welfare, education, and medicine.

Mainland authorities have now confirmed that Dr Ho's monopoly licence will continue beyond the return to China until its expiry at the end of 2001. After that, the future administration of the casinos is uncertain. Some people see the present spate of gang violence in Macao as part of a jostling for position of different criminal interest groups who hope to take advantage of the changes, and Dr Ho's eventual retirement, after the return to China.

Although Dr Ho is best known today for his rule of the gambling kingdom, he has also been involved in earlier chapters of Macao's history. Dr Ho fled to Macao as a penniless young man from Hong Kong, when the territory fell to invading Japanese forces in 1941, and was soon taken on by the Portuguese administration to trade with the Japanese. He made a small fortune and a network of personal contacts across Macao, which would lay the groundwork for his association with the enclave for the rest of his life. Later, the Portuguese administration frequently called on Dr Ho for advice during the period of civil unrest in 1966, and he gives his own account of the closed-door exchanges that took place at the time.

I BELONG TO ONE OF THE WEALTHIEST FAMILIES in Hong Kong. My grand uncle was Sir Robert Hotung [the prominent and influential Hong Kong businessman] and my grandfather was the Honourable Ho-fok. He was a member of the Legislative Council and Compradore of Jardine's [one of Hong Kong's largest trading companies] for some years. My father was a director of the Hong Kong trading company, Sassoons, so we were very, very wealthy; however, when I was thirteen years old, my father speculated heavily in Jardine's shares, with his father and his brother, and they lost heavily. He was on the verge of bankruptcy, and he ran away to Saigon.

Until then, I had been a poor student. At thirteen, my mother told me: 'I can't support your studying any more unless you get a scholarship.' So, out of the blue, I became a very good student and managed to get a scholarship to class 3D, the worst class. From then on I got a scholarship every year to class 2A and 1A until I went to the University of Hong Kong, also on a scholarship. There, I studied arts for a few months but I didn't find it very challenging—there were so few classes, just here and there, morning and afternoon, so I switched to science and studied until the third year, when the war broke out.

I joined the air-raid warden's office and we fought with the Japanese for about two weeks, until the Hong Kong government surrendered. I had to throw away my uniform. Being a student, I had no money at all. Even at university, I was completely broke. I only had ten dollars in my pocket, my pay for the first ten days.

I had an uncle in Macao who called me and said: 'What is the point of staying in Hong Kong now? You'd better come over here and help me.' So I went, and he recommended me to the biggest company in Macao during the war, the Macau Cooperative Company Limited (MCC). It was one-third owned by Dr Lobo, the director of economics, one-third owned by the wealthiest families of Macao, and one-third owned by the Japanese army. I became its secretary because of my knowledge of chemistry, and because they knew I could be trusted, being the grand-nephew of Sir Robert Hotung.

The company's objective was to provide food for Macao during those three years and eight months of war. I had to start by learning Japanese and Portuguese, because my job was to barter between the two. The

Portuguese government supplied us with all the surplus they could afford to give away—tug boats, launches, telephone equipment, anything they could part with—and I exchanged all that with the Japanese authorities, in the name of the company, for food from the Mainland. We supplied flour and rice, beans, oil, sugar, all the necessities to support Macao because the Portuguese government wasn't very wealthy and they had to get all these supplies from the Mainland.

Macao was paradise during the war. In Hong Kong, many people, even some of my relatives, suffered considerably because of shortage of food, bombings, and harassment from the Japanese gendarmes, but the Japanese honoured the neutrality of Macao and they didn't interfere with the administration in any way. Their only involvement was in supplying food to Macao. In those days, if you had money, you could enjoy the best kind of cigarettes, American, British, right up to the end of the war. If you had money, you could carry on using motorcars and motorbikes all through the war—gasoline was available. And you could have excellent food—if you had the money. I had big parties almost every night. Bird's nests, roast pork . . .

Agreed, there was a lot of killing. I became the teacher of the most important Japanese man in Macao during the war. There was a Japanese Special Branch in Macao, which was even more important than the Japanese Consul General. The head of it, a man called Colonel Sawa, went to see the governor one day and told him he wanted to learn English—but he needed a reliable person who mustn't murder him. The governor thought about it and said: 'What do you think of the grand-nephew of Sir Robert Hotung? You were such good friends with Sir Robert—would you trust a member of his family?'

Colonel Sawa accepted immediately. From then on, he sent his car—with no number plate, just one Japanese star—to my house to pick me up, every morning at six, and drove me to Zhongshan, across the border in China. There, the two of us would climb together to the top of a small hill. Then he started singing in Japanese and taught me how to sing with him—and in return I taught him English. I was his teacher for one year, and in that time all the Japanese soldiers in Special Branch would kneel down to him—and to me, as his teacher. What a great difference! While my relatives, my mother, were suffering in Hong Kong, the Japanese gave me excellent treatment.

Macao only suffered one day of bombing by the Americans—amazing! One morning, we were very surprised to see planes flying over Macao, and the bombing started. It went on for about twenty minutes. Their target was the gasoline depot in the outer harbour. The Americans made sure it was blown up very quickly. Then we saw the American planes dropping mines all along the inner and outer harbour channels. They wanted to make sure any Japanese patrol boats using these channels would be blown up. They succeeded. As a result of the bombing, there was no more transportation between Hong Kong and Macao. A few passenger boats were blown to pieces, Japanese patrol boats were blown to pieces. Only a few small junks managed to get through because they were so small, they didn't trigger the mines.

Macao had the biggest population in its history during the war, over half a million people in so small a place. In those days, there were no developments in Taipa and Coloane, nothing. Everyone was crammed in this tiny little spot, Macao. While the war was on, there was no construction. We just made the best possible use of the existing old buildings. Now and then, we saw some Portuguese going to the outer harbour to board a battleship and go back to Lisbon. I think that happened three or four times. The funniest part was always Portuguese National Day when we saw the British and Japanese consuls general racing to see who could shake hands with the governor first.

There was only one month of starvation, near the end of the war, when the price of rice became too high. I couldn't get enough supplies from the Japanese. I risked my life several times. Even when the channels were mined, I took a trip to Guangzhou with a load of brass shells from cannons—that was all the Portuguese government had left to give away. So I took them to Guangzhou and made a very good barter for three loads of rice. I still remember when I came back into the inner harbour channel—all the people cheering, clapping their hands, setting off firecrackers to welcome my return. That was dangerous. In that month, people starved. We had people lying dead on the road.

The casino did excellent business. It was well patronized by the Japanese, and they lost a lot of money to the casino. Of course there were prostitutes; in those days, it was legal. One of the back streets was full of brothels, and they did a roaring trade. The restaurants did well. There was no unemployment. Everyone could find a job. Macao was booming.

At the beginning of the war, I had raised the question: 'Why should I work with the Japanese when I've already seen how they treat people in Hong Kong?' The governor said: 'This is an order from the Portuguese government. You and three others are appointed to deal with food. Food is important. Without food, the Macao people will starve. I have already written to the Chinese government to say that when they come in, they can't harass you, and can't do anything to harm you, because you are under orders from the Portuguese government.' So four of us were given special exemption papers by the governor. I still have that paper.

When the Portuguese government saw the Japanese were losing and the war was coming to an end, they formed a government bureau and I became a semi-government official in charge of the supplies division, a superintendent. The job was basically the same, it just meant I was getting supplies through the government bureau instead of the MCC. When I was with the bureau, I started a small trading company myself. By the end of the war, I'd earned over a million dollars—having started with just ten.

There were only two casinos at that time, one big one in Hotel Centro, the other one in a private shop, in the red light district. They were both open 24 hours. The small gambling den was next door to an opium shop, where they provided the gamblers with opium. All the croupiers were in slippers or wooden clogs. It was well patronized. Even in the Hotel Centro, where the premises were better equipped, there was no air conditioning, just fans. That was the one the Japanese had gone to during the war.

In the hotel casino, the staff were a bit better dressed, in Chinese pyjamas, not singlets. They only had three games—Fan-tan, dice, and Pai Kao, a sort of Chinese dominoes—no roulette. It was always full of people. Even my mother, after I managed to get her to Macao, loved to gamble there. They did a roaring business and made a lot of money. Macao was tiny, and yet a bit like Casablanca—all the secret intelligence, the murders, the gambling—it was a very exciting place.

When the war ended, I bought a boat, the first one to start crossing between Hong Kong and Macao, so I was in the shipping business and also in textiles. After two more years in Macao, I went back to Hong Kong and started trading with the Department of Supplies, Trading, and Industry. I bought up many of the supplies left behind by the British army.

I was a refugee during the war, on the run. I could have been caught by the Japanese at any time. If any one of my enemies had pointed me out to the Japanese and said: 'Stanley Ho was in uniform—he fought against you for two weeks', I'd have been thrown straight into Stanley Prison in Hong Kong. The Portuguese people treated me well, they were very kind to me.

It was quite a contrast when I came back to Hong Kong after the war. Having been born and bred in Hong Kong, I thought I knew the British. When I first started business, I knocked at the door of one of the senior British officials. I could hardly hear his voice, it was so low. I had to put my ear against the door. Barely audible: 'Come in.' I walked in and tried to shake hands with him but he was so cold, he wouldn't even shake my hand, just told me to sit down and said: 'What can I do for you?' I said: 'I have an export licence here for approval, I want to ship something to Macao.' He took it from me and said: 'I'll deal with it later. I'll let you know in due course.' That's dealing with the English.

In Macao, after the war, I started trading. I went to the equivalent department, to one commander in charge, a Portuguese. I knocked at the door and put my ear against it, expecting the same treatment as I got in Hong Kong. I almost jumped out of my skin when he bellowed: 'Come in!' Inside, I went to shake hands and he pushed my hand aside and grabbed me in a big hug. He said: 'Look! I'm so busy!' Piles of paper everywhere. I gave him the export licence. He grabbed my paper, wrote 'Already approved' across it and gave it back to me. What a difference! I've learned a lot of patience dealing with the Portuguese, also a lot of understanding dealing with the English. The Portuguese are more friendly. But I must admit, for administration, the British are the best.

I carried on in business for some years and then switched to real estate and did some business with Hongkong Land. Because of my time in Macao, I knew everyone there and in 1961, when there was a tender for the gaming concession, I joined with some friends and made a bid for the casino—and we won the franchise. I was in business with Mr Henry Fok, my brother-in-law Teddy Yip, and Yip Hon, who was a professional gambler. I needed someone who knew that side of the business.

I bid for the franchise for one reason. I was a Hong Kong boy and saw Hong Kong doing so well after the war, I saw no reason why Macao couldn't do the same. I made important promises to make sure we won the bid. In those days, you could only go from Hong Kong to Macao on the slow boat to China, you couldn't go on a day trip. It took five hours or more. You had to go all the way into the inner harbour. All the hotels were sub-standard, and so were the casinos, and they only offered Chinese games. So I promised in my tender: first, that I'd try to bring Macao nearer to Hong Kong by opening up a new channel and providing high-speed ferries. Second, I promised to build high-class hotels. Third, I promised to introduce European games into the casinos, and to make sure the croupiers were properly dressed and well educated.

I also promised to clear the resettlements along the Praia Grande and outer harbour. In those days, the early 1960s, all these Nationalist farmers, who'd run away from Guangzhou and the Pearl River Delta to settle in Macao, occupied the outer harbour. There were over a thousand families in little squatter huts. One of my obligations was to clear all that away. I also promised to dredge and maintain the new channel I opened, and the old channel, for the years of my franchise.

I think I've fulfilled all my promises, to the surprise of many people in Macao. They said Stanley Ho was a dreamer, he made empty promises. Now they can see I meant every word I said. We were lucky. Everything went my way. More tourists went to Macao. When I first took over the franchise, we had less than half a million tourists going to Macao, including those from Hong Kong. In 1996, we had more than eight million visitors. For a place with a small population of 450,000, if you have eight million visitors, that's really something.

One of the toughest times was the political crisis in 1966. All my casino directors deserted me. They were so afraid, they went back to Hong Kong. I was the only one holding the fort. I remember very clearly what happened. The Chinese confronted the Portuguese authorities to show their support for the Cultural Revolution. They believed in Chairman Mao's Little Red Book. Despite good treatment by the Portuguese authorities, they were still not satisfied, and managed to find some small matter to blow the whole thing up and turn it into a riot.

They picked on the head of the education department, about his dealings with a Catholic school and a Communist school. They accused him of giving every facility to the Catholic school and virtually nothing to the

Communist school. They rounded up all the students at the Communist school and staged a big demonstration along the streets of Taipa and that started it all off. They ended up forming a struggle committee, made up of the women's union, the labour union, the schools union, the commercial union, the sports union . . . thirteen organizations altogether. Mr Ho Yin was one of the struggle committee leaders. They went to the governor and demanded a strong, official apology to be issued in Portuguese and Chinese through the radio and the press, and said the education department must treat the Communist school in the same way as the Catholic school from now on.

It was very hard for the government to admit it had done anything wrong. After all, it hadn't—this whole thing was just an excuse to create commotion. Unfortunately, this all happened during the time of an acting governor. His time was up and he returned to Portugal, leaving a new governor to cope with it all. When the new governor, Brigadier Nobre de Carvalho arrived, I welcomed him in Hong Kong and took him across to Macao. A few days later, he invited me to the Palace. Things were already quite tense. He said: 'Mr Stanley Ho, I want your opinion. You are an important person in Macao and I've lined up Mr Ho Yin and yourself to hear your opinions.'

I said: 'You have to stand firm. If you give ground on this, no one will come to Macao any more.' He said: 'Yes, Stanley, I know. But the situation is getting very tense. They're beginning a riot.' I said: 'Well, you were given five demands. If I were you, I'd meet one or two of them and reject the three that are most difficult for the administration to accept.' He was on the verge of accepting my idea when my time was up and Mr Ho Yin arrived. Naturally, Mr Ho Yin, a member of the struggle committee, told him to surrender and accept all five demands. Before the governor had a chance to think about what we'd said, the riots were already starting. They were pulling down the statues, burning papers, burning down the Leal Senado, shouting in the streets. The governor didn't know what to do. After about one day of confrontation, he ordered a curfew.

The Portuguese soldiers in Macao had never experienced a curfew in their lives. Some of them wanted to show off. Eight innocent people were shot. Someone hung out of the windows to see what was going on—bang!—they were shot. Someone came out into the street to close up his shop more securely—bang!—shot. Eight people died. Then the matter became more serious. The struggle committee wanted compensation, a huge amount of money the government couldn't afford.

They started to refuse service to all the Portuguese. The Portuguese would go to the barber's shop and be told: 'No, you're Portuguese. Cut your own hair.' They went to the restaurants and the waiters refused to serve them—'You go home and eat.' That continued for about one month. I had to close the casino after two weeks because it was so tense. The taxi drivers also joined the union and they refused to drive all foreigners. Even tourists couldn't get a taxi if they came. I had no choice—I closed down for two weeks. Every day you could see the women from the women's union and the children from the Communist school, fourteen and fifteen years old, dancing in front of the Government Palace, and waving red and yellow flags. It was quite a sight.

During those two weeks when everything was closed, the governor called me almost every other day to discuss the situation. The poor governor! In his ashtray, there were about fifty cigarette butts. He couldn't work out a solution. At one stage, he wanted to pack up and go. He said: 'We're fed up. If they don't want us here, fine, we'll pack up and leave.' But it wasn't that easy. The struggle committee told the governor: 'Don't even think about leaving. We won't let you until you've paid the compensation and given us an apology.' They wanted the governor to sign a document saying the government killed eight innocent Chinese on purpose, something like that.

I told the governor: 'These are demonstrations to support the Cultural Revolution, but the Chinese central government may not want you to leave. They're not ready. You can't occupy Macao for 435 years and leave just like that. Maybe you should consult your government in Portugal.' And that's what he did. Finally Colonel Barros, Minister of Defence and a former governor of Macao, came. He was a good friend of mine. He came and saw two people, me and Mr Ho Yin. We had to meet him in a secret place so no one could see us. Not in the Palace, not in a public restaurant. I found him a very tiny hotel opposite the Peak Hospital. I rented a suite and we went there for several hours to discuss it all.

The colonel told me: 'I want a firm, firm opinion. Don't give me anything else. I won't listen to surrender.' I said: 'No, on the contrary, even when I saw the governor, I told him to stand firm. Why should you surrender? You don't need to. But, unfortunately, I have to ask you to tell the governor to agree to sign this unfortunate document under duress. It doesn't matter. The whole world will know he's signing it under duress. Let them use whatever words they want to. Let them say your Portuguese administration killed eight innocent Chinese. Tell him to close his eyes and sign. After that, I think the central government wants you to carry on.'

He said: 'But what if there is more harassment?' I replied: 'I doubt it. They don't want Macao to carry on suffering the way we are at the moment. I have no income from the casino. You have no income from gaming tax.' In fact, no one was paying any kind of tax because the struggle committee ordered all citizens of Macao to stop paying anything to the government until they got an apology and they signed this document.

I said: 'I really suggest this time you agree to both conditions. After that, Macao can return to normal.' Then he saw Mr Ho Yin, who naturally asked him to comply with the demands and told him Macao would prosper again if he did. The following day the governor, instead of signing this document in the Palace, had to go to the Chinese Chamber of Commerce and sign it there; and that was the end of the confrontation.

I wasn't worried at all. I knew it was a bluff from the start. I didn't believe that China, if it wanted the return of Macao, would go about it this way. After the success of the riots in Macao, they tried the same thing in Hong Kong. People came here from Hong Kong to learn from the struggle committee, but in Hong Kong, they all went to jail because the English stood firm and arrested all of them. The British won completely. In Macao, the government surrendered—but the Church won.

During the confrontation, Mr Ho Yin was appointed to go and see the Portuguese bishop, on behalf of the struggle committee. He took with him ten copies of Chairman Mao's Little Red Book. It was during the Chinese New Year holidays. He said: 'After New Year, you must teach Mao's thoughts in all your Catholic schools.' The bishop said: 'This is a very serious matter. You can't expect a reply from me just like that. I must contact the Vatican and consult the Pope. I'll give you a reply as soon as possible.' Mr Ho Yin said: 'Not too long—we can't afford to wait.' Within forty-eight hours, the bishop received a telex from the Pope saying: 'Close all the churches, close all your schools if necessary. Please reply accordingly to the struggle committee.'

So the bishop called Mr Ho Yin and showed him the message. 'Here's my reply. I can translate the telex into Chinese for you if you like. You can't take it away with you but you can write down what it says and go back and tell your committee.' Ho Yin took the message back to the committee and there was lots of furious shouting. They ruined the walls of the churches, all the Catholic schools, daubed them with all kinds of slogans—Victory to Mao! And all that sort of thing. The main doors of the churches were kept closed. The schools were still closed because the holidays weren't yet over. But when the holidays did end, all the Catholic schools stayed closed. They outnumbered the Communist schools by far; the Church ran about 70 per cent of the schools. So all the kids were wandering round with nothing to do.

The New China News Agency, which was controlling things behind the scenes, called an urgent meeting because all the families were saying: 'What are we supposed to do? We've got to go to work. We haven't got time to look after our kids now that they're not going to school any more.' The Communist schools didn't have room to take all the children themselves. So they held a big meeting with lots of furious arguing, and in the end, the director of the New China News Agency gave the order: 'We have to agree with this bishop—tell him never mind about teaching Mao any more.'

Mr Ho Yin went back to the bishop and said: 'I've got good news for you. You can open all your schools again. Forget about teaching Mao's thoughts.' The bishop replied: 'Well, it's not that easy, Mr Ho Yin. First, you clean off all the slogans from the walls of the churches and the walls of the schools, and then I'll see what I can do.' Another stormy meeting of the New China News Agency took place, with everyone cursing the bishop up and down, and finally they surrendered. One morning, between three and five in the morning, they sent out all the workers to clean up all the schools and churches, and the Catholic church won. A 100 per cent victory.

Macao used to be a very peaceful place, a place people went to rest, unlike Hong Kong where in Central you hardly have to walk, people just push you along. Unfortunately, the security situation in Macao is not good any more, so we've had a drop in tourism and in the gaming business. Last year we had a drop of some 18 per cent. Even so, I've been in Macao for thirty-seven years and the tourism and gambling business is the core business, the most important pillar supporting the economy. The government now relies on the gaming tax for as much as 60 per cent of its revenue, which is very high for any government.

I was told that in 1996 the Hong Kong police, in order to give a clean city back to China, launched a tough anti-triad campaign and tried to drive them out of the territory. As usual, the Hong Kong police were very competent, and did clean up a lot. With such easy immigration laws in Macao, most of the triads came to Macao for shelter, but Macao is a tiny place—it cannot accommodate so many triads. In 1997, Taiwan did the same thing—the police cleaned up in Taipei. Again, many of the triads came to Macao.

Portugal recalled its soldiers in 1976, when it declared that Macao was a special territory and no longer a Portuguese colony. That was not a clever move. In Hong Kong, they kept soldiers until the very last day before the handover. Now, Portugal can only rely on the police. With the cost of living going up every day and the police working on a very tight budget, it has been suggested that some of the police, in order to earn more money and knowing that 1999 was coming up, may have become involved with the triads to earn more money. Since 1996, we've had this violence in Macao, and the government has found it very hard to rely totally on the police, because part of the force is really involved with the triads.

Most of the fights are gang fights, gang members fighting each other for a bigger slice of the cake. It's not a very big cake now—just the casinos, some gambling business, some smuggling. Fortunately, up to now, they have only targeted gang members or government servants controlling them, not tourists. I'm also told there's one other reason for all this. They say the police were too kind to some gangs and not so kind to other gangs and that again caused fights. I can't say any more about it. It's not my job. I'm a trader. It's up to the police to do something to keep the place safe. My police friends elsewhere tell me that for a place as small as Macao, it's difficult to have two police stations. There is the security police and the judicial police, and the two never cooperate; in fact, they're always fighting with each other. But it's too late now to change anything.

I think the Chinese were very pragmatic in dealing with Macao. They've already extended our gambling franchise beyond the handover. Now the franchise expiry date is 31 December 2001. What will happen to Macao in the future? Let me tell you. Macao has no natural resources. The industry is almost gone. I would say, in the last five years, at least 80 per cent of the industry has gone to the Mainland because labour is too expensive. In any case, the number of workers available in Macao is very small, less than a hundred thousand. A while ago there was a boom in real estate. Mainland people could get money easily from the banks and they invested blindly in Macao real estate. They were building heavily without demand. As a result, we have some 50,000 empty flats. So if the government can't rely on the sale of land, can't rely on industry, what else can it rely on? Really, the future is still tourism and gambling.

Gambling is one of the six evils in mainland law, but here in Macao it will be permitted to continue for fifty years. No other part of China will be permitted gambling, not even Hong Kong, but Macao's Basic Law, article 118, says the existing entertainment industry may carry on for another fifty years—which means we can carry on gambling. Not only is Hong Kong not permitted gambling, Hong Kong operators won't be allowed to take over gambling in Macao. It has to be locals. So if one day I lose the franchise, it will still be in the hands of local Macao people.

We are quite fortunate that, throughout this four hundred and fifty years of Portuguese rule, they managed to leave behind a heavy sense of European culture. They were clever to preserve all the heritage buildings, unlike Hong Kong. Tourists should still be interested in all these buildings, partly European, partly Chinese. And Macao is famous for its excellent food, in both Chinese and Portuguese restaurants. The tourists will still come after the handover.

David Chow Kam-fai

From Junkets to Politics

The main business of gambling in Macao's licensed casinos is controlled by Stanley Ho's company, STDM, but there are other businesses on the fringe of casino life that allow other businessmen to make profits indirectly from the vast gambling industry.

David Chow is a directly elected legislator who recently entered the Legislative Assembly for the first time. He has made much of his money by running junkets for the casino industry, marketing the casinos overseas, and arranging special promotional tours for those who want to come to Macao and gamble. The nature of such deals varies, but part of the business involves investigating would-be clients to make sure their credit is sound, arranging special, even free, deals on accommodation in local hotels, and providing other perks to big-spending customers. Destinations such as Taiwan and Japan have proved particularly lucrative, especially before the onset of the Asian financial crisis, although customers can be tempted to Macao's gaming tables from all over the world.

Mr Chow also has interests in real estate in Macao, and is developing his political career in the hope of providing a voice in the Legislative Assembly that can represent the needs of the gambling industry.

MY FATHER ORIGINALLY came from mainland China, then relocated to Hong Kong; my mother was from Macao. When they married, they settled in Hong Kong and my first memories are of being there. In those days, our family was well known and connected to a lot of famous people. We used to be a big name in real estate. My father had condominiums on Connaught Road. In the 1950s, when there were hardly any multi-storey buildings, he owned several.

We lived on MacDonnell Road, in a big apartment with lots of servants and drivers. We were very rich, but my family's fortunes rose and fell. In the late 1960s, after the riots in Hong Kong, property prices crashed and we lost everything. The property market fell 70 or 80 per cent because so many people left Hong Kong. We had a lot of real estate money invested in construction, and the bank came after us, chasing up on its loans. We went bankrupt. My mother struggled on for a while, but property prices didn't really recover. The bank put pressure on us and, in the end, we had to give up and leave.

I went to school in Hong Kong until I was sixteen. Then I had to leave and start work—we just didn't have enough money for me to carry on studying. I took all kinds of jobs, hard labouring jobs. I fixed televisions, worked as a delivery boy, worked in the banks as a clerk and auditor. Anything that made money, I'd do it. I needed money to survive. People looked down at us then. They had what the Chinese call 'pigeon eyes'—you know what they are?—eyes with dollar signs etched over them. We had been rich. Then, suddenly, we were poor. My mother had to go out to work too, washing dishes, cooking—she was the one who really kept us going.

Those difficult times taught me a lot. They taught me that sometimes money means everything, other times you have to work for your principles. I've seen the best of times and the worst of times. When you're rich, you can do what you like. Those times belong to the past, I don't want to discuss them. I've been rich and I've been poor—and that means I know what life's all about.

When I was twenty, I went to the United States and lived there, working and studying there for about eleven years. I started out in Los Angeles. For the first two years, I had some financial help from my family, especially my sister. After that, I had to pay for myself. I received a diploma in English and continued, slowly, for four years. I would travel and work to earn some money, then go back to school and study until my money ran out, then go travelling again. I worked in twenty-six states altogether. Finally, I ended up back in LA, went to college there, and graduated with a degree in hotel management.

By that time my mother had moved back to Macao and was managing a hotel, working for Stanley Ho. He was a good friend of the family before we went broke, but we stayed friends. My mother made a deal with Stanley, which meant she could manage a small hotel here, and she called me back too. I realized that in Chinese culture, even if you're educated, you need personal connections if you want to get ahead. So I came back in 1981 and have been in Macao ever since.

This was a small town when I returned, but I could tell Macao had a lot of potential. My real strength is tourist promotion, and at that time I could see great opportunities for tourism in the future. I'd had a lot of experience travelling, promoting the tourism business, and I had managed restaurants and hotels, so this seemed like the thing for me. In Macao, gambling is very important. We're a small place, we don't have many resources. Macao needs the help of the money from the casinos to develop other parts of the economy.

I became involved in the casino business, organizing what we call 'junkets'. It's a type of tourist promotion, similar to the system they use in Las Vegas, although you need a special licence to run them there. I'd worked there when I was in the United States, and I had built up good contacts, which really helped. In those days, the revenue from the casinos in Macao wasn't that good, so I developed a junket programme for STDM. Stanley Ho accepted it and we started up the business. We devised promotional tour programmes for the casinos, by putting together packages for people in other countries. STDM paid our commission based on what our visitors spent in the casinos. Our job was to solve problems, make sure things went smoothly, and make sure the visitors to Macao were happy.

At that time, the Portuguese didn't promote Macao hard enough. They needed help from local people like me who speak the casino language. Lots of places in the region—like Australia, the Philippines—didn't have a casino business of their own, so there was plenty of potential. Good marketing was crucial. We attracted business from lots of countries— Korea, Taiwan, Malaysia, and of course Hong Kong. Macao is in a great location, with easy access to China, just a short flight away from most other Asian countries, and only an hour away from Hong Kong by jetfoil. When I started, I hired one girl and five others in the team. By 1985, I founded another company with Singaporean and Indonesian partners, then I began the most successful years of my life.

We put Macao on the gambling map. People were coming on our trips from all over the region, all over the world. I know how to communicate, how to make friends. I have connections with the gambling industry everywhere. I was bringing in hundreds of people a month. I'm still working hard, trying to develop Macao as a destination. I think we have more opportunities than any other corner of Asia.

Gambling itself isn't a bad thing—but you've got to control it. We don't ask people to come here and spend all their money and bankrupt themselves. Everyone has their own habits. Some people like beautiful watches, some like beautiful cars, some like beautiful jewellery. Everyone has their tastes, and gambling is the most enjoyable entertainment in the world, especially for Asian people. Sometimes I gamble myself—just for fun. It's easier to communicate with other people when you're sitting at the same table and enjoying the same thing. But you have to have self-control. I can lose and it doesn't mean a thing to me. The point is whether you take it too seriously. Just like the stock market and the business world—everything in life is a gamble. It depends how you handle the risks. In Monaco, the royal family owns the licence for the casinos. I think they also enjoy gambling. You go to clubs in London, people enjoy gambling. Macao was never designed to attract hardened gamblers. It's for tourists—but they don't lose too much. We want to develop a good image as a place for gambling, for the benefit of the whole of society.

We have our rules. To start with, people have to bring their own cash, cash only. Sometimes, if they lose a little bit more, they'll ask for some credit—all over the world, in America too, the casinos have a credit facility. Rich people like to be trusted. If you go a few times and you pay well, we have references, which we check out just like any bank would do, it will help us to issue credit. You win or lose. If you lose, you go and sign a cheque, and pay us back. That's part of the marketing job. The credit department is the biggest department in any casino. It's like a bank—it's where you issue the money for the casino. You have to know the business to work there. It's not something you can teach. You have to feel it.

In Europe, high-society people like to gamble. The poor don't have the money to enjoy it. Gambling is a business, and you have to make sure people enjoy themselves. When you start a casino, you have a lot of overhead expenses. You need money to support the casino, but the important thing is how you administer and control it. Look at Las Vegas. Before, the Mafia controlled Las Vegas; now, all their gambling industry is corporately owned. This is how the casino industry brings benefit to the society—it's got to be balanced. If all the money is going into one pocket, gambling is a bad thing. It's the atmosphere that counts—that's one of the most important points. Good or bad, it's how the place feels.

Everywhere in the world, there are good people and bad people. A good man and a bad man—what's the difference? You think all businessmen are good men? You think all government officials are good men? You think all police officers are good men? You have good and bad people all over the world, but it's how you distinguish between them that matters. If someone is doing something bad, you have to stop and think: was it because you gave them the opportunity to behave badly? If someone goes to jail, even juveniles, don't condemn them at first; give them a chance and re-educate them.

We have to make the casino business here more noble. We have to make Macao a place people want to visit, not only for gambling. We don't want gamblers gambling, we want tourists gambling. You spend a few days here, and spend a thousand patacas [Macao's currency, roughly equivalent in value to the Hong Kong dollar] on a hotel room, five hundred on the slot machines; but if you pick a good number, you'll be very happy. If you just have a bit of money, maybe not enough to play at the tables, you can try your luck at the slot machines just for fun and spend some time there.

People should feel they can come here to relax and enjoy themselves for a couple of days, not just to gamble. We need modern management, we need to be corporatized, we need to show people Macao is really somewhere. We have to move forward—we're years behind Hong Kong. If I could do one thing, I'd like to attract international investment here, and then really put Macao on the map. But some journalists in Hong Kong do so much damage to Macao's image—a motorcycle or car is set on fire and it's splashed all across the front pages.

At the moment, things are so political because the Portuguese are going. This is the hardest period in the history of Macao because of the politics. Why should we rely on the Portuguese? They're leaving. Would you depend on someone leaving his job? The Macanese [Eurasian] families are different. They've been here for centuries, their roots are here so they want to stay, of course they do. And that's good. If everyone here is Chinese, we're just the same as Zhuhai [a special economic zone], just across the border. We want to keep Macao international, so I support all foreigners. I support the Macanese people and their right to stay here.

I've got my own interests in the industry and I want to see things improve. How do we make it better? There's no point in Macao people fighting each other for control of the casinos—that's pointless. We need to face up to our competition and do better than other places to make sure we get international investment. We need better city planning. Look at Las Vegas—the atmosphere there is much better than in Macao. We've got to make changes. We have more than seven million people coming here every year. If we doubled that number—and there's no reason why we couldn't, we've got the potential—we wouldn't have to fight each other, there would be plenty for everyone. We've got a great climate. We're a multi-lingual place, with four hundred and fifty years of European culture behind us. We've got more historic buildings than Hong Kong. So why are we so far behind Hong Kong?

I became involved in politics three years ago because there isn't a single person representing the gambling sector. I think people have to understand why the Portuguese don't care about the casino industry— all they care about is the monopoly licence and the huge revenue they get. When they hand Macao over to China, things probably will be different. China has no gambling industry, so how are we going to promote gambling in Macao? Gambling is the major revenue earner here, so who is going to explain to politicians in Beijing about the industry? I'm trying to do something for society by spreading the word myself. I try to communicate with people, and try to work something out for myself by broadcasting Macao's message to the rest of the world. I'm a tourism guy. I've built a career out of junkets and marketing. I'm trying to do something to help and I thought entering politics will probably help Macao in this field.

In the near future, the higher levels of the Chinese government will pay more attention to Macao. Macao isn't going to be a big problem because economically we have less than half a million people, no stock market, and people here are very steady, not as greedy as in Hong Kong. Macao profits a lot from the casino industry. Our income per capita is seventeenth in the world. People here drive Mercedes Benz— in Beijing, they drive Audis.

In the last ten to fifteen years, money from the Mainland has really started to move in here. Many corporations are now owned by the Chinese, and when Macao experienced its real estate boom in the late 1980s, 80 per cent of the money came from mainland China. They've made a big investment here. Macao is different from Hong Kong. You know what the British left behind when they left Hong Kong? The new airport and huge financial reserves. But what happened after the handover? Hong Kong's economy crashed because of the Asian crisis, and all the resources went to Shanghai. Big investors prefer Shanghai to Hong Kong, and that's the reason Hong Kong is suffering; but China still owns 80 per cent of the property here in Macao, so the future scenario for us is far happier.

The casino industry is also the way to attract people from Europe to invest in Macao. We should develop a big resort here, something like in Las Vegas. It wouldn't be very difficult because labour here is cheap, materials are cheap, everything is cheaper than in Hong Kong. I have a plan to build one for about US$200 million. In the US, it would cost about US$650 million for the same project. In Las Vegas, the annual income is US$7 billion. Here in Macao, without the same facilities as they have, our annual income is already about US$2 billion. If we really worked at it, we could be even more successful than Las Vegas. Travellers love to have a change—they don't want to go to Las Vegas every year. So Macao has a chance; after all, we have the culture the Americans don't have.

I'm sure that, after 1999, the Chinese government will give the authorities in Macao the necessary autonomy and will respect the Joint Declaration, and that the people of Macao will govern Macao for another fifty years.

Kung Yick

The Gambling General

When I met him, Mr Kung was eighty-six years old and lived in an senior citizens' home in Macao. He began his life in Anhui Province and spent his early career as a soldier, fighting first against the invading Japanese forces and later in China's civil war where he fought against the Communists as part of the Kuomintang (KMT or Nationalist) forces. He fled from China when the KMT were defeated, passing through Hong Kong before settling in Macao.

Kung Yick left China a very rich man. As well as his own wealth, he was also entrusted with a large amount of money controlled by the Nationalists. Some of it he gave to fleeing Nationalists as they passed through Hong Kong to Taiwan. The rest he took with him to Macao, where he started a lifetime of gambling addiction in the casinos.

He lost that initial fortune, and a lifetime's earnings, through gambling, mostly in the casinos but also at the dog races. He died in December 1998, penniless, but with no regrets for his gambling sprees. In this account, he gives some insight into the compulsion to gamble—which was at the centre of his life—and which makes Macao's casino industry such a thriving and profitable business.

I STARTED LIFE AS A SOLDIER in China. I enrolled in China's military academy in 1935, then later trained to be a military advisor. I joined the Chinese army in 1937 and fought for the next eight years against the Japanese. When the war finally ended in 1947, I thought it was all over—but then the civil war started. I fought in that war for two years, fighting against the Communists on the side of the Kuomintang.

I was very depressed towards the end. I wanted to devote myself to the party, to the Nationalist cause, but it was clear ordinary people in China were ready for revolution and supported the Communists. It was hard for me to accept—I was deeply disappointed in them. I fled

China in 1949 and went to Hong Kong with my wife and two children, a boy and a girl.

When I arrived I was very rich. The Nationalists had paid me well, and also had given me a lot of money to take out of China to help people escaping from the Mainland. I had a total of about US$3 million—it was a fortune in those days. I used just over a million of that to help people who were fleeing through Hong Kong to Taiwan and needed to set up there. But I still had more than a million dollars left for myself. You could have bought the whole of Macao with that! The original plan was to go to Taiwan with my wife and children and start a new life there. But my wife and I had a huge quarrel—she didn't like Hong Kong and wanted to go back to China. She left me and went back to China with the children because she said she liked life there. As soon as she set foot on Chinese soil again, the Communists seized her and she was thrown into prison. Eventually they sent her to some remote part of northern China, to some wasteland.

She wrote to me asking for help and I stayed in Hong Kong for several years, until 1953, hoping to find a way of getting the family out to join me, hoping there was something the Nationalists could do to help us. But we couldn't do anything. In those few years in Hong Kong, I ran through a lot of my money and decided to leave and come to Macao to build a life here. I was young—I wanted to have some fun. As soon as I hit Macao, I spent all my money. I never heard from my wife and children again.

When I arrived in Macao, I thought it was great. Life here was cheap and free. The streets were calm, people didn't seem nervous. It was a good life—lots of entertainment, lots of beautiful girls, lots of fun. That's how I spent all my money, on prostitutes and gambling. There was a priest in St Augustine Church who was working on a new dictionary and I went there to help him. The pay was US$50 a month, which was a lot in those days, so I worked there for three years until the dictionary was finished and the priest went back to the United States.

Then I got a job riding a bicycle rickshaw, taking local people and tourists round Macao. The pay wasn't bad, the equivalent today of about US$1,300 a month. So I had a good income from then on. I had my own bicycle rickshaw and wasn't bothered by anyone, the gangs left

me alone. There were two communities in Macao then, one pro-
Communist and one pro-Nationalist, and we Nationalists used to come
together every now and then for meetings. But when the 1 2 3 uprising
happened in the 1960s, the governments in Macao and Beijing signed
an agreement to say that all Nationalists had to leave Macao, so the
associations run by the Nationalists were closed down and most of
them moved to Taiwan. I was pretty much left alone here in Macao. I
carried on working as a bicycle rickshaw rider right up until the 1990s.
It was a good living.

When I started as a rickshaw rider in the 1960s, I earned lots of money.
I bought a private villa for myself—I was rich. But I was already gambling
a lot and by the late 1960s, I started to run into money problems. I had
to sell the villa and from then on, I slept in the rickshaw at night, I
couldn't afford anywhere else. I spent everything in the casinos. I started
gambling as soon as I arrived in Macao—from then on, I went almost
every night for more than forty years.

I liked to gamble on dice, in a Chinese dice game called Dai Siu. You
roll the dice and bet on the size of the numbers. When I first came, the
Lisboa Casino hadn't even opened. I used to go to the Estoril Casino,
Macao's first casino, and later to the Hotel Centro. When the Lisboa
opened, and the Floating Restaurant as well, I started going there, too.
I usually went in the evening. I would finish work, have dinner, and
then go along on my own. If I'd had a good day and earned a lot, I
would go to the casinos. If I hadn't done too well, I would go to the dog
races. You don't need as much money for the dogs.

I wouldn't say it was particularly exciting. The only thrill was if it
was a really big stake—if I played hundreds of dollars all at once. I
knew everyone at the casinos, all the staff, ladies, boys, everyone. They
used to call me 'the General', that was my nickname because I was
from the military. The casinos were always busy. People flocked there
from all over the world; overseas Chinese, too. I was a famous guy in
those days. Even if I won a lot, I could leave the casino safely because
they all knew me and knew I was a regular customer. No one would
threaten me or make me feel uncomfortable.

Forty years ago, when I started gambling, there were no criminals.
You could win, walk out, and feel safe. But in recent years, the gangs
are much more involved. If they know you're a foreigner, they follow

you out of the casinso—especially if you've won a lot—wait until you get to a quiet place and rob you. They're all gang members.

I'm not very interested in money. The only point of money is to buy entertainment. I'm a pretty depressed person, and gambling is a way of cheering myself up. It's a habit. I don't go to get rich, the money isn't the issue. I just don't have anything else to do. If I lost heavily one day, I would still go the next day—why not?

I don't have any regrets—I like gambling. If you gamble, you're going to lose sometimes—that's the rule of the game. I'd say in the course of my life I must have lost about five billion patacas in the casinos. I've got nothing now. But I've no regrets.

I think it's right for Macao to have the casinos. They're a big money earner for the government. If they all closed down, government revenues would really suffer. If someone turned round now, right now, and gave me some money . . . if it was a really big sum, say thousands or millions of dollars, I might keep it for something else, but if it was less than that, I'd go straight to the casinos and bet it.

Former Triad

Gangs and Gangsters

Getting an insider's account of membership in a Macao triad society is no easy task. When they join, members swear a lifetime allegiance to a triad brotherhood—those who break the rules or talk to outsiders about triad society business run the risk of putting themselves in a position of personal danger.

This man, now middle-aged, was recruited to the Macao branch of a triad society when still at school. He was finally expelled from the society after many years of membership and direct activity in triad criminal activities. His own drug addiction had become uncontrollable, increased by repeated spells in Macao's only prison. His addiction made him less useful to the society's criminal activities and he was dropped.

His recovery began after he met a Macao social worker who helped him to go off drugs and start a new life, away from Macao's criminal gangs. He now works full time with a local charity. He was unwilling to be identified for fear of reprisals from previous triad contacts.

The triad societies are international Chinese crime syndicates, roughly equivalent to the Mafia. They are thought to have started as secret societies that opposed corrupt rule in mainland China in the nineteenth century. Nowadays their main areas of activity are smuggling, drug dealing, money laundering, prostitution, and loan sharking. This makes the casino-dominated economy of Macao an attractive destination; several of the larger international triad societies, such as the 14K, compete for a share of the illegal profits made on the fringes of the gambling industry in Macao.

Although the Macao branches of these societies have local hierarchies and leaders, officials in Macao often point to the fact that their activities are on an international scale—with connections in Hong Kong, southern China, and Taiwan, as well as elsewhere in the world—and have to be tackled on an international scale. One suggestion is that it was the crackdown on triad activities in Hong Kong in the 1980s and early 1990s that contributed to an increase in triad activities in Macao, as criminals came under pressure in Hong Kong and relocated.

This man alleges that triad membership is widespread in Macao, at a grassroots level. He estimates that some 30,000 people out of a population of 450,000 have allegiance to one society or another. This doesn't mean that all are engaged actively in illegal triad activities; most have token membership in a society and, for example, pay its local leaders protection money to ensure a rival society does not damage their family or business interests. Security officials in Macao put triad membership at a far lower level.

I WAS BORN IN BURMA but had to leave as a teenager, during the political upheavals of 1972, when they started expelling the ethnic Chinese. I fled to Macao with my older brothers. I went to school but I only stayed for about a year and a half. My brothers started working as soon as we arrived.

I didn't do any work at school, I just messed about and became involved with the gangs. I was still young so I wasn't a proper gang member; we just hung out together, went to parties, and had fun. I fell in with a bad crowd, and when we finally dropped out of school, we all joined the triads together.

We went through an initiation ceremony before we were accepted into the society. We had to kneel down and pay our respects to Guandi, the god worshipped by the triad societies. I wanted excitement, something to do. In the mornings and afternoons we slept, and then met up in the early evenings to go out together. To start with, we just fooled around, doing nothing much really. Afterwards, when we got a bit older, I became involved in illegal rackets to make money.

I've got tattoos—this one on my chest is an eagle. I did it myself, in prison. The warders gave me needles and I did it because there wasn't much else to do. But tattoos weren't compulsory in the gang. Members of the same triad society recognize each other by a gesture—whenever you meet another member, you have to do this action with your hands to show you're brothers. The triad society I joined was one of the biggest in Macao. There must have been about 20,000 members altogether, and I knew many of them.

I did a lot of smuggling. No one actually organized us, but because we were part of the gang, we knew what was going on. We'd know when the ships were coming in and exactly what we had to do. We smuggled many people, illegal immigrants trying to get out of China, and some domestic appliances like television sets, hi-fis, radios, that sort of thing. Later, I started smuggling drugs.

When I was working with the gang, I never took drugs. That came later. I got caught and sent to prison for two years, and started doing drugs inside. It was easy to get supplies in prison; we got them from other inmates and the warders were taking money from us too. It was out of boredom as much as anything—there wasn't anything to do inside.

I didn't notice the impact of heroin at all when I first started injecting. I felt fine—it didn't affect my health. But when I came out of prison, I realized I was addicted. I couldn't live without heroin. I needed to carry on getting it, and that meant I needed a lot of money to keep myself supplied. I went back to the gangs, but this time I was just selling drugs. The drugs came to us through Hong Kong; I don't know where they came from originally. This was the 1980s and I was making about US$1,300 a month—or more—smuggling drugs. Everything I earned, I spent on heroin for myself. I was pretty happy with my life then.

Gang life was like having a load of brothers. People really got on well together and looked out for each other. If you were in the same triad society, there was never any trouble, but if you met members of rival triad societies, there were fights. Sometimes groups of us would live together, renting a room or a flat and all staying there. Other times, I rented a place on my own. This life carried on much the same for about twelve years. I was in and out of prison all this time—I must have served about five or six sentences.

Once, when out of prison, I tried to join up with the gang, but they wouldn't let me near them. They thought I was useless and didn't want anything to do with me. Even when I was out of prison I couldn't work any more because I was so addicted to heroin, so they abandoned me. I was really upset and angry. I made up my mind, I wouldn't even try to follow them, I'd look after myself from then on.

I had no way of making money. I stole and did a few illegal things, got caught, ended up back in prison again. I was in and out of prison and not making any money at all. I was getting into a worse and worse state, and ended up sleeping in the street or in derelict buildings. The last time I came out of prison, in 1992, I didn't know what to do. Someone introduced me to some social workers and that was the turning point for me. I went to meet people working with the charity, Caritas. Although they couldn't solve my problems straightaway, they gave me money to live on and I slowly began to change my way of thinking.

It was a long, hard battle coming off heroin. I went through treatment programmes seven or eight times. As soon as I finished the programmes, I'd go looking for heroin again. But I did it. I'm in good health now.

When I first joined the triads, they weren't really that bad. I'd say out of half a million people in Macao, about thirty thousand of them were in one triad society or another. It wasn't seen as such a bad thing. So when I joined, my brothers weren't too worried. People joined because they wanted a sense of belonging, a sense of shelter. People liked to gather together, and feel they were all part of the same group and had some protection.

After the initiation ceremony, there weren't any special meetings or rituals. We'd just gather together informally at certain restaurants. Nothing special. If you wanted to leave, you had to get permission from the gang leaders. If you were really serious about leaving, you had to make sure you had someone on your side looking out for you; once you'd left the gang, you might get trouble from the other members. You needed an official protecting you, or someone high up in the police, or else you would be in danger.

There were girls in the triads too. They were mostly wives or girlfriends of members. Sometimes they didn't actually join themselves, they'd hang out with the men. But some could join properly and have the same initiation rites. Then, if they had an argument and broke up with their boyfriend, they could still stay in the gang in their own right.

The way I see it, it isn't the triads that have caused trouble in Macao, it's the government. That's what needs sorting out, not the criminal gangs. I've experienced it myself—the bribery and corruption. If you want to do something illegal, and it's looking hard to achieve, you just give more money and bribe your way through. I think it might be better after the handover. Things have definitely become worse because of the increasing number of illegal immigrants in Macao. Four or five illegal immigrants join together and call themselves a new gang. That's why there are so many people involved in crime now, and they're all rivalling each other for money.

My real brothers have all left Macao. They emigrated to Taiwan or countries in South-East Asia. As for the gang members, I never see them. I don't want anything to do with them now. No way. If I could have my life again, I'd do everything differently. I wouldn't live it the way I did again.

Brigadier Manuel Monge

Fighting Organized Crime

Brigadier Monge is the Secretary for Security in Macao. It has proved a difficult post in the run-up to the handover, a period in which media coverage of Macao has been dominated by the dramatic outbreaks of gang-related violence, thought to be related to the casino industry.

Brigadier Monge first came to Macao in July 1987, just a few months after China and Portugal signed the Joint Declaration. The President sent him to the enclave to test public opinion and provide feedback to Lisbon at a critical political time. By coincidence, the Brigadier already knew one of the people on his list of interviewees, Dr Jorge Rangel: the two had met years earlier when they had both served in the Portuguese army in Guinea-Bissau.

In 1991, the Brigadier was named as the President's special advisor on Macao and later, in September 1996, took up the post of Secretary for Security. For the first time, he was based in the enclave. Here he gives his thoughts on the violence that has rocked Macao, and its causes, and responds to the allegations of corruption made against the two local security forces.

SOME THINGS SEEM SO DIFFERENT now compared with my first visit in 1987. I remember going to Zhuhai, just across the border; at that time, there was almost nothing there. The horse and cart was the standard method of transport. Other things haven't changed. For example, I remember meeting the man who was the head of security in Macao at the time, and he had exactly the same furniture in his office as I have now! The military is very conservative.

In a relatively short time, most things have changed dramatically in Zhuhai and Macao. My memory of that time was of anxiety, of uncertainty about the political changes ahead and what they would all mean for Macao. I was impressed with the many young people I met here. They had strong views about the transition, and seemed to know exactly what they wanted for the future. They accepted that Macao would return to China, but they also felt strongly that it must keep its basic identity and characteristics. This was one of the most striking memories I have of that time, this sense of conviction amongst the territory's young generation of business people and academics.

The governor of the day, and other senior administrators, focused on the need to develop opportunities for these young people. The government was already involved in negotiations with the owners of the university here—they wanted to buy it and turn it into a public institution so they could initiate new courses and offer young people more career choices. In fact, that did happen not long after my visit. In 1989, the government took over the administration of the university. We started new schools within it, for law, engineering, and so on. I was also asked to recommend to Lisbon the development of other academic institutes to broaden the range of opportunities for young people. The head of the security forces in Macao also urged me to start a training school here for new recruits. In 1988 we founded the academy for training members of the police and security forces.

I was aware that Macao was in an unusual situation. The political changes here, the devolution of power from Lisbon to Macao, and the gradual introduction of democracy, hadn't come about as a result of the local population demanding more say in the running of their daily affairs. In fact, they were the result of Portugal's own political changes in 1974, with the development of democracy there.

In 1989, the President visited Macao for the first time, and I also came along as his advisor. The President wanted to walk around the streets anonymously, so he could see for himself what was going on. One day a few of us wandered into a small Chinese restaurant where no one spoke a word of Portuguese. Of course, we couldn't speak Cantonese either and we had to struggle to communicate just by pointing out what we wanted to eat and using sign language. I really enjoyed the experience and was impressed by how kind the people were.

The following day I told the President about it and he was very curious and eager to try this restaurant himself. So I returned and said I wanted to book a table for dinner, because I'd enjoyed it so much the day before. We didn't tell them it was the President who was coming.

Unfortunately the President got sick and couldn't go—but the rest of us did and we were really surprised because, just as a kind gesture, they'd made a number of changes and decorated the restaurant beautifully to greet us. To this day, they don't know it was the President who had been planning to eat there. I was impressed that the fact we couldn't speak the same language did not stop them from giving us a special welcome.

In 1991 the President named me his special advisor on Macao, the only advisor in the Presidential Palace in Lisbon working specifically on the enclave. Lisbon is a long way from Macao, and the President felt he needed someone to keep regular contact with the government here, so he always knew exactly what was going on.

There had already been a couple of serious incidents. In March 1990, Macao faced an immigration crisis when some 40,000 immigrants from the Mainland staged a demonstration here. Two days after that, there were signs of the beginning of a strike in the police force. I flew out then, and intervened to make sure stability was guaranteed.

In September 1996, I was appointed Secretary for Security, and came to live in Macao. Until 1990, the commander of the security forces had been at the same level as an under-secretary in the administration, and had a seat in the executive branch of government. But the post of Secretary for Security was created only in 1990. As the special advisor on Macao, I had already had a great deal of contact with the key people working on security issues, so there weren't any big surprises when I first arrived. The only real surprise was being appointed to the job by the President in the first place!

I was aware when I took the job that it would be a difficult one. This is a time of change and there are bound to be complications. We could already see signs of the Asian financial crisis, and economic problems always bring an increase in crime. But it was a challenge I wanted to accept.

When we talk about crime, we have to distinguish between two categories of crime—organized crime and common crime. Common crime is often the result of economic problems. When there is an economic crisis, rising unemployment, a fall in local incomes, then of course there is an increase in petty crimes, robberies and so on. This happens everywhere and we've seen it all over Asia.

The other problem is organized crime. First of all, one has to appreciate that the triads have been around for centuries. Their presence and motivations in the early days were quite different from today. During the Japanese occupation of China, the triads played an important underground role fighting against the Japanese. Later, at the time of the Communist revolution, the triads were dormant. They couldn't develop and were suffocated by the strength of the revolution. With the opening up of Chinese society, especially in the Special Economic Zones, they had a new lease on life, becoming involved in illegal aspects of the new business opportunities opening up.

In Hong Kong, the triads flourished in the 1970s and 1980s. As a result, Hong Kong's anti-corruption body, the Independent Commission Against Corruption (ICAC), had to strengthen itself to fight the triads who were becoming involved in illegal big business, like smuggling. In Macao, the triad problem is much the same. Macao has had triads for a long time and their strength varies depending on the strength of the local economy. In the late 1980s, we saw a tremendous economic boom in Macao, and that led to a growth in illegal businesses. There was a lot of money to be made. People were investing heavily here, in construction and real estate, and a huge number of joint ventures were set up between investors from mainland China and local people—some of whom were also involved with the triads. Of course it's easy to analyse this now, in hindsight.

The situation changed in 1993, when the rate of investment from the Mainland began to slow down. The Chinese authorities became concerned about the flow of capital out of the country, and they started to tighten their control. The triads were already heavily involved in various businesses here including, of course, the gambling industry, which offers plenty of opportunities for crime. Macao soon began to have economic problems. Some businesses were in crisis because investments had been made, which were proving unsound. The same thing, of course, was happening all over Asia in the second half of the 1990s.

In 1997, the authorities in China started to wage an effective battle against organized crime, and against the triads in particular. On 2 July 1998, when President Jiang Zemin was visiting Hong Kong, he declared that the triads, and organized crime in general, were a great threat to

China. The Chinese authorities launched a special operation, mainly in southern China, against organized crime and activities in which the triads were heavily involved, such as drug trafficking and smuggling.

When we talk about the triads, we're talking about cross-border crime. There's no such thing as triads who are only from Macao. Macao has divisions of organized crime syndicates who set up here, but they are also strong in Hong Kong and Guangdong province, where the 14K was first founded. This is an important point, because Macao is an open territory with transparent borders. Macao is a small place with a large gambling industry, and that attracts triads. It's a trend, not unlike the crime wave that hit Casablanca decades ago, or Chicago. It's a problem— but it is not confined to Macao.

People don't talk very much about crime in mainland China, but they have a major problem there too. The papers report that every year thousands of people are executed on the Mainland because the authorities are trying to clamp down on criminal activities. Because Macao is so transparent, any criminal activities here are given a high profile.

In the last three years, we have seen a turf war between rival triad gangs. Our analysis is that there are two gangs jockeying for a share of illegal activities, especially those related to gambling. We are coming up to the handover, the gambling licence expires in the year 2001 and has to be renewed, and certain people are trying to make sure they are in a strong position when that happens. Also, Stanley Ho is getting older and it's more difficult for him to keep control of all the activities. There are some powerful people who think control of the gambling industry should not be concentrated in just one company.

All these factors have contributed to the problem—both the political situation and the economic one too—but we should keep this in perspective. We're talking about twenty-nine homicides in the whole of 1997, and twenty-seven in 1998. We think about two-thirds of these cases are related to in-fighting between organized crime gangs. That gives us a crime rate of 4.5 homicides per 100,000 inhabitants, which is the same as the world average—and much lower than many cities, such as Washington.

That's not an excuse—I'm just making the point that Macao is not a dangerous society. If this didn't involve the triads, we wouldn't even be talking about crime. But let's face it, it's not exactly normal to see the leader of a triad gang—and not even the most powerful triad gang in Macao—giving personal interviews to important international magazines like *Time* or *Newsweek*. People have a special curiosity about gambling and the triads. It's a product the media sells very well. We accept we've got a problem, and we have to deal with it, but it's ridiculous to imply Macao is some sort of war zone and local people are dodging bullets every day. That's just not realistic.

The Minister of Security of Cape Verde came to see me recently and showed me his crime rates; the number of homicides there is higher than in Macao, but you never hear about them having a dreadful crime problem. The crime situation here in Macao is seized on by the media and relayed to people all over the world—but the same level of crime in other parts of the world wouldn't even make it into the newspapers.

Portugal and China are still negotiating about Macao and there are various issues still to be worked out. Each side will try to take advantage of the other's problems. That's natural. It happens in any process of negotiation. That lies in the political arena. As far as operations are concerned, China is cooperating with us. The Chinese authorities are well aware of the magnitude of the problem, and of the fact that it is not just Macao's problem, it's a cross-border one.

So many people cross that border every day. In our one prison in Macao, half the criminals are not Macao residents. Macao has a lot of imported crime. China and Macao have cooperated well in the last year or so. China was aware it had to take stronger control of the border between Macao and China, which has traditionally been mostly the Mainland's concern. When it comes to political disagreement, we understand the situation and we have to accept some criticism, but when it comes to ground operations, we have improved cooperation with China in all our efforts to fight organized crime. That's why, since October 1997, China has increased the number of special police at the border. They have more than 2,000 special officers working there now.

I wouldn't say I'm fully satisfied with our fight against organized crime—or with China's efforts—because my goal is a 100 per cent success rate. We can't be complacent. As far as China's co-operation with us is concerned, I would say I'm satisfied. We've had positive responses to our requests for help, particularly for help at the border in looking for criminals who are trying to escape to the Mainland, and vice versa.

As for the issue of corruption in the police force, I'll say this: there's infiltration in every organization. Even in some governments with strong democratic traditions, like in the United Kingdom and Germany, you still occasionally find people, even at the ministerial level, who are corrupt. Of course we can't say there are no bad elements anywhere in our police forces. There are people in police forces all over the world who are tempted by crime and criminal organizations. That's the reality. It's the same in Macao.

But it's quite untrue and unfair to say that the whole Macao police force is infiltrated by the triads, or that the triads somehow dominate the police force. That's just not true. We have good disciplinary procedures. In 1997, for example, we dismissed thirty-six officers. Not all of them were accused of triad involvement, but they were all found guilty of behaviour not of a sufficiently high standard for them to stay in the police. In 1998, we dismissed thirty officers. At the moment, there are twelve cases pending which we expect to lead to more dismissals. It's a tough battle against corrupt elements.

There are ways of showing that our ability to operate isn't compromised by these bad elements. Our undercover operations aren't damaged by information leaks. We also have a number of police officers responsible for the personal security of key people in Macao. Most of them are Chinese police officers, and if we didn't trust them, we wouldn't chose them for such sensitive jobs.

Glamourizing the triads is a real problem. Some of the blame has to go to the Hong Kong film industry, which is very strong and influential, and, according to police intelligence, mostly in the hands of the triads. Of course their films reflect their values. Most of the films show violence, organized crime gangs, and their activities. This message helps to form the values of the young people who watch these films. The importance of money, expensive cars, beautiful girls—these are the values that, unfortunately, are emphasized by most of the films produced by this industry—along with violence. Recently, we set up a special task force to focus on the problem of juvenile delinquency. At these meetings, a main concern of teachers and youth leaders was violence, and the way these movies influence young people. This is something we have to fight.

European and American magazines also glamourize the activities of the triads and this turf war between the different gangs in Macao. It's part of a fascination with the Far East, and with the criminal underworld of the triads. They're portrayed as mysterious, exotic, and exciting, but that's actually a very colonial attitude. It's a way of looking at the Far East as a distant and intriguing place, but one that has no connection to their own lives and situation. One day, the problem of organized crime will touch their societies—and the day it does, they'll change their attitude. Organized crime, including drug trafficking and other underground business activities, will be one of the biggest problems to confront the world in the next century.

I'd advise other countries to start seeing this as their problem too, instead of glamourizing it. They might not be affected by it now—but I'm sure they will be, because these organized groups are looking for opportunities all over the world.

Culture at the Crossroads

As Macao faces political changes, after more than four centuries of Portuguese administration, more subtle social changes are also taking place, with profound implications for Macao's future identity. A debate is underway: will Macao rapidly turn its back on its past links with Europe in embracing its new Chinese identity, or will it keep its distinctive East-West blend?

In the final years before the return to China, the Portuguese administration laid the groundwork for preserving its legacy in a number of areas; for example, by lovingly restoring Macao's neo-classical architecture.

Figures such as Gary Ngai have championed the need for Macao to remain culturally distinct, from both mainland China and Hong Kong. Its special identity, they argue, a mix of Portuguese and Chinese heritage and language, will be its future strength and the cornerstone of its competitiveness. Macao is sometimes described as an unpolished diamond. The challenge is for the first post-handover government to polish it, not let it be overshadowed by Hong Kong or the growing economic strength of Zhuhai, the Special Economic Zone just across the border in mainland China.

The Macanese community—some 10,000 local people with mixed-race heritage—often have a strong Portuguese influence in their lifestyle, as well as elements taken from their Chinese background. They are among those most determined to preserve the traces of Europe. If Macao loses its historical legacy, the enclave will become just another Chinese city, they say, indistinguishable from other mainland cities, and with no special attractions for investment, or tourists. Their own interests, too, may be at stake. In the past, the Macanese community thrived in its role as a bridge between Portuguese and Chinese communities. If the Portuguese influence wanes, will that role become redundant? Some Macanese don't speak fluent Cantonese and fear feeling estranged from the place of their birth if Portuguese is subsumed.

While some people fervently defend Macao's multicultural traits, and see their preservation as essential to its future, others are more sceptical. Some point to the population profile. Nearly 30 per cent of the present population came to Macao from mainland China in the ten to fifteen years before the handover. Many have little contact with the Portuguese and Macanese communities in daily life, and retain a strong sense of their Chinese cultural identity. More than 96 per cent of the population use Cantonese as their mother tongue—while less than 2 per cent use Portuguese. Gary Ngai and Carlos Marreiros defend Macao's dual cultural heritage and discuss the reasons why it should be preserved, while an immigrant from mainland China, Mrs Chen and her young son, who came to Macao illegally to join her husband, give their impressions of life in Macao and the imminent return to Chinese administration.

One significant social change, evident in recent decades, is the gradual move on land of Macao's fishing communities. One estimate is that in the period between World War I and World War II, one-third of the local population worked in the fishing industry. Now, that number has shrunk dramatically to less than three-quarters of 1 per cent of the population.

The nature of fishing has change, too. Traditional methods involved junk-based coastal fishing. A whole family, often three generations, lived together on one boat, which was both home and workplace. Members of a fishing family were born on their boats, married on them, gave birth on them, and died on them.

The relationship with the land people was chequered. The fishing community was seen as an inferior minority group, and its members were ostracized by those on land. The way of life made formal schooling impossible, and most were illiterate. It is only in the last decade that members of the fishing community have begun to integrate, to receive a more stable education, and to intermarry in Macao, although most members of the fishing community, even today, prefer to chose husbands and wives from within the community.

Fishing had been important for Macao for centuries, since before the Portuguese sailors and merchants settled there; however, this century has proved a turning point for the industry that has steadily declined and is now facing virtual extinction.

During World War II, the local fishing fleet fell from 12,000 boats at the start of the war, to just 4,000 at its end. The most significant changes, however, came after the war. Until the end of the 1960s, all local fishing followed traditional lines, using sailing junks, hemp nets made and mended by hand, and there was no mechanization. In the early 1970s, the process of modernization began with a vengeance. Nylon nets were introduced, and the junks were gradually converted and became motorized. In Hong Kong, the government played an active part in assisting the modernization process, but in Macao the story is rather different.

The Macao fishermen were left largely to their own devices, and an informal system of loans developed within the industry. Fishermen, needing capital to modernize, were forced to turn to the U-Lans, the powerful middle companies who bought fish wholesale from the fishermen and then sold it to restaurants and markets at a profit. In taking on these loans, the fishermen were forced to sell their catch exclusively to a particular U-Lan, while they were paying off their debts. This indebtedness gave the U-Lans a great deal of power in determining prices, making it more difficult for fishing families to clear their debt.

More recently, land reclamation and an increase in marine pollution, including the disruption caused by the construction of Macao's new international airport, added to the fishing community's woes, making it increasingly difficult for them to earn a living on the traditional fishing grounds. By the 1980s, many families had abandoned their junks, moving ashore and looking for new work, often in the construction industry, which was booming at the time.

The traditional family unit, which lived and worked together on the boat, has almost completely disappeared. Those who do still fish are crews made up of mostly male relatives, supplemented by some hired hands, on much larger vessels, that travel to deeper waters, on longer trips. This means some families are divided, with older people and children left behind on land.

Macao's fishing community is of special interest because it has faced relatively little disruption until recently, compared with Hong Kong fishing communities that underwent rapid and intense modernization, and mainland fishing communities whose traditional way of life was affected by the Cultural Revolution in the 1960s. Macao's fishing community reached the present decade with its own distinct rituals and festivals preserved intact.

Photograph courtesy of the Historical Archives of Macau

Gary Ngai

Macao's Cultural Future

G ary Ngai, a well-known commentator on Macao's culture, has passionate views about Macao's future potential and cultural heritage. He is the executive director of the newly formed Macau Sino-Latin Foundation, a private body that aims to promote closer ties between Macao and other Latin-language countries and, through Macao, between those countries and China. During twenty years in the enclave, he has worked as a senior broadcaster, civil servant, and advisor to the governor during the sensitive time of Lisbon's negotiations with Beijing on Macao's future.

His initial career in broadcasting included the post of assistant director at Radio Macau, where he was in charge of the Chinese language programmes. In the early 1980s, he was a leading figure in the establishment of Macau Television, Macao's first television station set up by the government—it has since become a government–private sector joint venture. In 1993, he was appointed vice-president of the Institute of Culture of Macau, holding the post until he retired from public life in 1997, at which time he took on his current position with the Macao Sino-Latin Foundation.

Gary Ngai was born in Indonesia, of Chinese descent. He left Jakarta for mainland China, after finishing high school in 1950, one of the first overseas Chinese from Indonesia to return. He was eager to contribute to the building of the newly formed People's Republic of China. After graduating from the People's University in Beijing, he worked as a translator for official foreign visitors, narrowly escaping the anti-rightist purge of the late 1950s. He says he was persecuted during the Cultural Revolution, the time when he takes up his story here.

WE SUFFERED A LOT during the Cultural Revolution. Those of us who had come from overseas and spoke foreign languages were accused of being imperialist agents. I was denounced and sent to a labour camp with my wife and children, who were just six and four then. Life was very, very hard. We were sent to north-east China, near the oil fields in Manchuria, and it was –40°C. When the Sino-Soviet border clash broke out, we were moved to the south of China, to one of its poorest areas. For eight months of the year, we had nothing to eat except dried pieces of sweet potato.

My mother-in-law left Indonesia because of the anti-Chinese movement there, but she became stranded in Macao with no one to look after her. In 1978, when China first started to open its doors, we all slipped out and settled in Macao. When I first arrived, I gave private language lessons, mostly in Mandarin and English. I also wrote articles for magazines and newspapers in Hong Kong about China.

Compared to Hong Kong, the media here is very negative. Before the Portuguese revolution, there was very strong media censorship. Neither the Portuguese nor Chinese press had freedom of speech. After 1974, the Portuguese press started to open up and the Portuguese newspapers began to speak out much more. On the Chinese side, the opposite happened. They were controlled by the so-called 'invisible hand'. You don't see this in Hong Kong but here it's very strong. The Chinese press is under the control of the Communist Party and you have to follow their lead. There just isn't the lively debate you get in Hong Kong. If you want to write something controversial, you go to Hong Kong and get it published there. Of course we get a lot of Hong Kong newspapers here and listen to Hong Kong radio and television; you could say it's a sort of intellectual export and re-import trade.

Journalists have always been subject to violent threats. That was happening twenty years ago when I arrived. It was particularly bad during the Tiananmen Square incident in 1989. People who tried to speak out against the crackdown were terrorized, even physically attacked, and beaten up by the triads. The Communist Party uses the triads to suppress so-called dissident voices. Either side can use the triads—as long as you pay. They don't have any political affiliation—they just want money.

We were always being interfered with when I worked in radio and television. Someone from Xinhua [the state-run New China News Agency] or the pro-Beijing wing would telephone and say: 'You shouldn't say that—calm down.' It was a friendly warning rather a threat, but we knew exactly what they meant. Sometimes we'd change a report, sometimes we wouldn't. We were a government-funded station, so the Chinese could only interfere so much.

Of course, if we were critical of the government, the government would tell us to stop, and then we really did have to do as we were told or we would have been fired the next day. That's why freedom of speech in Macao is so difficult. Both sides interfere. I didn't like it—but I had become used to it in China. I could adapt if I had to. If I couldn't broadcast something, I could always write an article for a newspaper in Hong Kong instead. The knowledge that we could be attacked for trying to speak out is deep-rooted in people who come from China. We were always under pressure. We could never speak out.

Some time after I had left the media to join the civil service, I was offered a job in the Governor's office, where I became an advisor on Chinese affairs. This was when the negotiations between the British and the Chinese on Hong Kong's future had started, and we wanted to begin our own negotiations on Macao. I supplied information about China and suggestions on how to negotiate with the Chinese. I made quite a contribution, but secretly, from behind closed doors.

Much of my work was about the transition—how to implement policies, the localization process, changes to the legal system, and making Chinese an official language in the territory. At that time Portuguese was the only official language. Chinese was recognized officially only in 1991, twenty years later than in Hong Kong. I travelled back and forth between Macao, Hong Kong, and Singapore to learn about the localization process and bring back advice for Macao.

In many ways, the political issues here are very different from those in Hong Kong. Under British rule, Hong Kong developed an influential middle class in the 1970s and 1980s, when the economy grew rapidly, and that middle class, mostly educated in the West, became independent. Economically and intellectually, they were quite separate both from China and the United Kingdom, and now they make up the mainstay of Hong Kong's political structure.

In Macao, we don't have this sort of middle class. Macao is economically underdeveloped. It hasn't grown to become a financial or trading centre on the scale of Hong Kong. Macao's economy has been in decline since the Opium Wars; we're a forgotten corner of the world, in the shadow of Hong Kong. Also, Macao is hindered by its size. When students graduate from high school, they go abroad and never return, because there are no jobs for them here. There is no space for them to develop their talents. The economy isn't advanced enough to absorb them.

Without a large, professional middle class, Macao's autonomy will suffer. It is already happening. The Preparatory Committee [the committee that developed the political foundations for the return of Macao to China] is controlled by the Communist Party, the underground Communist Party, which, in effect, took control here in 1966 after the 1 2 3 incident. The Portuguese have been a lame-duck government ever since then. They can't do anything without the approval of the so-called shadow government here, Xinhua. The Governor has no real power. Chris Patten [the last British Governor in Hong Kong] could do anything he wanted but in Macao, the Governor has to consult the Chinese before he makes any major move.

Sometimes that Portuguese habit of bowing to Chinese pressure is very destructive. Take foreign investment, for example. Macao is a free port, one of the only two free ports in China along with Hong Kong. We need to do all we can to attract more foreign investment, but because of this weak government, and the strong conservative influence of the Communist Party, it hasn't been allowed to happen. Foreign investment here is only 4 per cent of the total—the rest comes from China and Hong Kong. That is ridiculously low. Macao isn't an international city in the way Hong Kong is. Everything here is under the control of the Chinese authorities.

Also, there's a big difference between the British and the Portuguese. The British became economically powerful in Hong Kong. Look at all the British companies that are major players in Hong Kong's economy. Britain wanted to protect its interests—it had strong economic reasons for doing so. That's not the case with the Portuguese. Although the Portuguese have been here for four and a half centuries, their interests in the economy are negligible—just a few Portuguese banks. That's it. In 1974, when the Portuguese revolution started, the Portuguese tried to give Macao back to China. 'We don't want Macao any more. It's too far away, very expensive—it has nothing to do with us any more.' But Mao Zedong and Zhou Enlai said: 'No—stay and look after it for us.'

China's strategy for Hong Kong and Macao was that they must serve as China's windows on the outside world. At that time, Western countries shunned China. It only had these two windows, and it had to keep them open. Their argument was: 'Let the British and the Portuguese carry on ruling them—it's to our benefit.' And they were right. That's the thinking behind the one country, two systems concept—Macao and Hong Kong will serve China better if those windows are kept open.

Of course, although Portugal's negotiations with China on Macao were relatively smooth compared with the negotiations over Hong Kong, there were sticking points. One was the future of the Macanese here, an issue Hong Kong didn't have to worry about. In Hong Kong, Eurasians were a small number and were discriminated against under the British. But here it's different. The Macanese are the middle men. Without them, the Portuguese couldn't survive because they're the only group that is bilingual. The Portuguese never spoke Chinese and they relied completely on the Macanese to explain to them what was going on.

Another sticking point was the issue of Macao's cultural heritage, whether Portuguese traditions should be preserved or not. This is a fight that continues. The Portuguese are afraid they will lose their heritage. Macao's culture is a tremendous resource, it is unique in the world. The Sino-Latin culture could be an important part of our future tourism.

They're now building this Ocean Park, a major theme park, and there's a big debate going on about what sort of park we should have. Some people say: 'Build something like Ocean Park in Hong Kong or Shenzhen. A mini-China.' I say: No—we don't want to duplicate what other places already have. We must develop something of our own, reflecting our Sino-Latin culture, emphasizing the differences with the Mainland. Old Chinese traditions, from the worship of the Chinese goddess A-Ma, to Taoism, and Buddhism, are still very much part of people's daily lives here. On the Mainland, much of that was destroyed in the Cultural Revolution. We can capitalize on the Portuguese legacy. Every year we spend millions of dollars preserving the old fortresses, the old churches. Some people say it's a waste of money—why not pull them down and make a lot of money building skyscrapers? That's so short-sighted. We have to fight that mentality hard, and preserve what we have.

A lot depends on the future leadership. If they recognize the value of this cultural heritage, it will survive, but it's all in the balance at the moment. We don't really know who will rule Macao after the handover. If it's a leadership that identifies itself with Socialist China, with Communist China, they'll destroy everything and build high-rise buildings.

My vision for Macao is that it becomes a bridge between the Latin world—in Europe, Africa, and of course Latin America—and China. We're talking about one-sixth of the world's population, a tremendous and largely underdeveloped market. We have special access to those countries because of our common law, culture, and language. It's a different family from Hong Kong and the Anglo-Saxon world. The question will be whether or not China sees Macao and its Latin identity as something precious, something that distinguishes it from the other cities in China—or whether it destroys it. If it kills it, it will be the end of Macao. We will just be a small district of Zhuhai.

At the moment, we're trying to emphasize to the young generation that they should learn not only English but Portuguese too. When I first came here twenty years ago, almost no Chinese, beyond a handful of civil servants, spoke Portuguese. There's a saying in Chinese: at the point where the tips of Macao's hills drop out of view, as you travel towards Hong Kong, that's the limit of the Portuguese language. The attitude used to be: 'Why should I bother to learn Portuguese? It's useless.' Now it's different. A lot of young people are signing up for Portuguese courses and evening classes—they're all fully booked. I'm delighted about that—it's a sign of a high civic consciousness, and the fact the young generation here want their own identity, something distinct from Hong Kong and mainland China.

Of course there's some resentment against the Portuguese here. The Portuguese have always discriminated against the Chinese, especially the police force. When I first came, I saw with my own eyes how they used to beat the Chinese. There was no equality. That was the main reason for the anti-Portuguese riots in 1966. Local people were furious—and they had a right to be.

Also, the corruption within the police and some parts of the civil service is dreadful. The Chinese won't tolerate it any more. Look at the security situation, with all the recent bombings and arson attacks. The Portuguese have done nothing to improve law and order. With so much corruption, it's impossible to reform public life, but people here shouldn't confuse culture and colonialism. They are two completely different things.

The problems with law and order are all about money and the economic downturn. The fastest way to get money is to kidnap someone. We have to crack down on these illegal activities. Macao is small. The border is transparent, and vulnerable to infiltration by illegal elements from Hong Kong and from the Mainland that try to make money here. It's such a small place, they inevitably run into each other. That can be stopped after the handover. If China, Hong Kong, and even Taiwan join together to make a coordinated effort, then it could work. Start with an anti-corruption campaign, clean up the police the way they did in Hong Kong in the 1960s. We can do the same in Macao. The key to it all is transparency. If everyone in Macao knew how to do things properly, and not through corruption, people would have more confidence in the system. At the moment, people know they have to give bribes along the way.

Macao's casinos are a big source of income, of course, but that doesn't mean they have to be a source of crime. Look at Las Vegas, Monte Carlo—they're clean. You can make a casino city a decent, safe place, if you have the right controls. It's a question of management. If the police and security forces worked closely with the casinos, they could sort it out.

Macao hasn't seen the waves of pre-handover emigration that Hong Kong did. Some 100,000 people in the present population have Portuguese passports, so they're not afraid of having nowhere to go, no emergency exit. They can go anywhere they like in Europe. There are some rich people who've moved their property overseas but not many.

The two separate communities in Macao are quite a phenomenon. The Portuguese were laissez-faire. In Hong Kong, the British forced the Chinese to learn English. Here, the Portuguese didn't learn Chinese, and the Chinese weren't forced to learn Portuguese, so there was no communication between the two races. The only bridge, for a long time, was the Macanese. The others didn't mix with each other. Now that's starting to change. You can see it in the statistics. The number of intermarriages between the two communities has increased dramatically. My own niece is married to a Portuguese man. There is a growing tolerance of culture on both sides.

Carlos Marreiros
Architect for the Future

Carlos Marreiros is a prominent figure in the Macanese community, with a strong interest in Macao's cultural heritage. As one of Macao's best-known architects, he has contributed to the urban development of the enclave, and has strong opinions about the city's physical legacy. He is also an accomplished artist—one of his paintings is used for the cover of this book. His work has often explored themes of Macao's cultural identity. He is now an active member of the organisation 'Macau Sempre', which is striving to maintain Macao's blend of Asian and European culture in such areas as language, architecture, cuisine, and literature.

Mr Marreiros is Macanese, with a mixed Portuguese and Chinese heritage. His father was born in Macao, but of Portuguese extraction. His mother was Eurasian, with a Portuguese father and a Chinese mother from the Mainland. As a result, Mr Marreiros grew up as a Catholic, speaking both Cantonese and Portuguese. He describes his typically Macanese childhood, characterized by large family gatherings with the unique fusion cuisine of the Macanese community.

It was common for previous generations of Macanese who went to live overseas for their tertiary education to settle abroad. Mr Marreiros is a member of the generation that returned to Macao and contributed to its development in recent decades.

MY CHILDHOOD WAS VERY HAPPY. I was born in 1957, when Macao was not exactly a rich place, but a stable and peaceful one. The Japanese occupation of China and Hong Kong had already occurred. I grew up mainly with my grandparents. My grandfather on my mother's side had ten children so I was used to a big, traditional family. I always had lots of cousins around to play with. I remember traditional parties, eating Macanese food, playing with all my cousins. Most houses in those days had a garden, so we had plenty of room to play. I was lucky. School life was great too. The high school we attended was huge, with a beautiful playground. We had many friends and enjoyed ourselves.

Macao was still very Mediterranean. Living in a small peninsula surrounded by water, gave us the sense of a constant dialogue with the sea. The city itself always had a special relationship with the water. The roads were cobbled, there were beautiful houses, described, wrongly in my opinion, as colonial architecture. They were really a local version of neo-classical architecture. The Chinese district had its blue brick houses.

Macao was not a sophisticated place. There weren't many children's playgrounds but we invented our own games—talking, playing football, building dens and hideouts. As a child, I had links with both the Portuguese and Chinese communities. I think I belong to a generation, people like me now in our forties, whose contact with the Chinese community was much greater than that of our parents and grandparents.

The only real disturbance was the 1 2 3 incident in the 1960s, Macao's miniature version of the Cultural Revolution on the mainland. At the time, I had just begun middle school, form one. I was about nine years old. As a child, the event didn't touch me directly. I remember the adults being agitated about it all. Some of them were talking about leaving and going back to Portugal. For a while it was rather difficult and confused, politically and socially.

I remember being with one of my aunties in a barber's shop near the Leal Senado, when there were strikes going on, and the demonstrations started. My aunt grabbed me and said: 'Come on, Carlos—we're going home.' I said: 'No—I want to stay and see what's going on.' But she wouldn't let me. All the schools were closed at the time. When we got home, I climbed onto the roof of our building—we had a two-storey house in a good area of town—to see if I could see anything. We lived in a quiet residential area with no high-rise buildings, just villas and two-storey houses, so I couldn't see much.

Once we became teenagers, people started to go overseas. Some went to Portugal to study, as I did later. Some went to other countries. Some came back afterwards, and some didn't. No one talked about deep political problems. Macao was a very conservative society—it still is—but in those days it was very conservative.

I only became aware of social problems when I was about sixteen, when we started reading books and learning about injustices in our society. We used to go across to Hong Kong often, not for noble reasons, but to meet girlfriends and have fun. In 1971, we all had our hair long, tied up in pony tails, wore scruffy blue jeans, and went around on motorbikes. We looked like hippies, although of course we weren't really hippies. You have to be very brave to be a hippy, and we weren't. It was just the fashion of the day. We did get a fresh social perspective because of the influence of the hippies, and that was increased by Portugal's revolution in 1974. I was finishing school at the time, and had planned to go to the United States to study, but when I heard about the revolution, I was very excited. I wanted to see for myself what was happening in Portugal.

At the time, links between Macao and Portugal were ridiculously poor. We found out about the Portuguese revolution through Hong Kong radio, some time before the Portuguese newspapers reported it. It was partly because Portugal was an autocratic regime, and also because there were no faxes, no telexes, only telegrams, and the news was issued by the Portuguese news agency, ANOP. The Portuguese papers, however, arrived in Macao late. All the news coming out of Portugal was disastrous, and in Macao it was amplified because the society was so conservative. Macao was far away from the motherland, so anything that happened had very strong echoes here.

When I announced to my family I wanted to go to Portugal to study, my father said: 'You're crazy—you want to go now to Portugal now, with everything that's happening?' I insisted. I said I wanted to see the revolution. In the end, I couldn't go that year because the faculty of architecture at the university was closed because of all the disruption. I had to wait a year and a half in Macao.

Seven of us were in the same situation and we all got junior jobs in the post office. I worked in the post office during the day, and in the evenings I taught Portuguese to Chinese students at night school. All my students were older than I was. They were workers, and for me, getting to know them was very good exposure because I had never worked before.

My morning job was ordinary—selling stamps and approving papers— and I had plenty of time to read books. I was one of the main readers at Hotung Library. In those days they had many books in Portuguese, English, and French, and because I was given a formal introduction to the head of the library, they gave me permission to take books away with me. I could sit and read at work when we weren't busy.

One of my colleagues was an elderly Chinese man who did basic errands. He used to talk to me about Tang poetry, simple Chinese proverbs, and fairy tales. They made a big impression on me, and I learnt a lot from him. In my evening classes, I learnt a lot about the Chinese way of thinking. My students weren't from rich families. They were studying Portuguese so they could get jobs in the civil service, because the government paid more than the private sector then. That experience helped me to get a more realistic, mature perspective.

Until then, my understanding of Chinese culture had been largely theoretical, gained through my Chinese grandmother and my grandfather, who was Portuguese and a Catholic but an open and tolerant person. He was my mentor. It was a rare thing for a Catholic to visit a Buddhist temple, but he used to take me along and say: 'You're Catholic, Carlos, but you must respect other religions.' I remember him taking me to temples several times to teach me about Buddhism.

I went to Portugal to study, with a deep conviction that I would return to Macao. Of my classmates, I'd say about 90 per cent returned to Macao after their university education. That was a real contrast with the generation before us who didn't—but in their day, there were no opportunities in Macao to tempt them back. I felt we had to come back. If all Macanese came back to Macao, the community would be stronger. We had our own traditions to maintain, our own language, our own literature, which draws on both Portuguese and Chinese heritage, our own architecture, and cuisine. When people talk about Portuguese culture in Macao, they really mean Portuguese culture generated in a different environment, shared with another civilization as well.

My parents worried that I might become involved in politics in Portugal, and do no work. It was certainly an extraordinary time. I saw the transformation taking place in Portuguese society. Portugal was still a backward country then, after forty-five years of an autocratic regime. I didn't join a political party, but I did have a strong feeling that Portugal had to modernize. It was sad to see people flocking to Portugal from the former colonies. They were well educated and looking for new opportunities—but they had left their homes and their heritage behind.

Politically, there were mistakes with the revolution, no one would deny that, but the overall result was very important. I remember the poverty and social injustices in Portugal at the time. I realised that fraternity and solidarity were very real concepts. I remember the strikes and the police chasing us. People supported each other and were generous in spirit. I remember seeing the police beat up a pregnant woman. We intervened and the police chased us until our friends helped us escape.

It was very important for my cultural development to be outside Macao in those years. I travelled all over Europe, in both communist and capitalist countries. I witnessed the rise of the neo-Nazis in Germany and realized how dangerous fanaticism is, be it on the left or right. I was away for eight years, and received a degree in architecture in Lisbon. I also taught and studied as a post-graduate in Germany and Stockholm.

I see myself as Macanese—but then one has to ask: what does that mean? We could debate that all day. I see the Macanese as people brought up here, who identify with this way of life. Macao is small, but I'm very proud of its heritage, not only its architectural traditions. I'm proud of the bureaucratic tradition, the attitude of tolerance. The Macanese are a unique cocktail of ethnic backgrounds and this means we can fit in and survive anywhere.

Macao is a free society. It's no bed of roses, there are many things one could criticize, but as a place to live, it's fantastic. In the last four centuries, we have had good and bad governments, we have faced real and political typhoons, yet Macao has survived. Macao never had the benefit of a primary sector. There is no space for agriculture. We don't have other natural resources to export. We're isolated—how could Macao possibly have survived all this time?

But Macao has survived—partly with the help of the government, partly with the support of the Church, but mostly through the strength of its civil society. Macao had the first Western university in the Far East. Through Macao, many new ideas have reached China—scientific and cultural ideas. Think of the new ideas, since the time of the Renaissance in Europe, brought to this part of the world through Macao. The Jesuits from Portugal, Spain, and Germany travelled to China and took with them some of the most advanced science, culture, and art of the time, from painting to medicine, astronomy to pharmacy. Macao was the gateway for all that. Naturally Macao was itself enriched in the process.

Dr Sun Yat-sen was a revolutionary, a republican trying to pull down the old dynastic system and found a modern China. His contact with Western ideas occurred here, in Macao, where he was a doctor. There are documents saying Sun Yat-sen wasn't really accepted in the Chinese traditional hospitals because he was too radical. He wanted to introduce Western ideas and techniques and he introduced them to China through Macao.

Historians have always considered Macao to be a humanitarian place. In this century alone we have accepted refugees from Russia, China, and elsewhere. Macao was damaged by Hong Kong's rise. It lost its strategic position. It was only in the beginning of the 1970s that Macao really began to grow and prosper, and really develop as a modern society. This has nothing to do with government, it's mainly because of civil society, and the prosperity that developed across Asia.

One problem was formal education. The people of Macao, both Portuguese and Chinese, have a very high regard for education, but the institutional framework just wasn't there. I don't just mean for tertiary education, but also a framework for learning values that are not only materialistic, from art and music to the humanities. Economically, Macao grew a lot from the 1970s onwards. As this economic development occurred, the Mediterranean Macao, the Macao of the poetic China trade paintings, faded away. There are both political and technical reasons for that. If, in the 1970s, we had moved the urban modernization to Taipa Island, we could have saved a significant part of Macao. But it's useless saying that now, isn't it? I'm not one to cry about the past. We have a lovely city with beautiful houses, trees, lots of space, and so on. Other cities were like that, too, and have been destroyed. The important thing is to find a balance. I think Macao already has been virtually destroyed in terms of its urban heritage.

There isn't only European heritage here. Most of the European inspiration is Portuguese and there are other contributions as well— Italian, Spanish, and even British. Hong Kong and Shanghai also have had a big influence. There are architectural patterns here inspired by Chinese people from Shanghai, whose sense of architecture had been inspired by the West. Although we talk about Portuguese influence, even the churches aren't 100 per cent Portuguese. We can find much local Chinese influence, and that of a range of other peoples from Japan to Sri Lanka, both in the architecture and the cuisine of Macao. When we preserve these collective memories, we are helping a society to look at itself in a mirror.

People without a past can have neither a present nor a future. To build a future, one needs to balance materialistic and economic values with spiritual and cultural values. It's a cultivated approach—and it's the only way to ensure balanced progress. It's a way of putting people at the centre of development. If you only have materialistic development, you run the risk of sacrificing the interests of the next generation.

Protecting cultural heritage is very important. That doesn't mean mummifying it, it means giving it new life. For instance, as long as a church still has life, or a Chinese temple, leave them alone. But look at the fortress. It hasn't had a military purpose for more than a hundred years, so of course it makes sense to change it to a hotel, or gardens, or a youth centre, or whatever. If we just paint and rehabilitate a physical structure and don't give it real life, it will be worth nothing. These buildings must be used by people. Cultural heritage is a must, it has to be.

I hope the heritage Macao has evolved in the last four centuries will continue. The intermarriage of races—the Portuguese, Malaccan, and Japanese, who are generally thought to be the first mothers of the Macanese, to the Indian and the people from other parts of Asia—Macao is the product of this mixture of cultures. The intermarriage with the Chinese is thought to be something very recent, only in the last hundred or a hundred and fifty years. Let's keep on with it. It's a very positive mixture.

My generation won't be the last. My daughters are mixed-race Macanese too. My wife is Macanese, in fact she's more Chinese than I am. A lot of people in my generation have intermarried, Chinese with Portuguese. It will carry on. In theory, the Macanese may start to die out, if in the future there is more intermarriage with the Chinese. That's a natural course of history, and if this is the will of the people, that's the way it is. So be it.

I think Macao will prosper for as long as we can keep a sense of our own distinct identity. The Macanese only represent a tiny portion of

Macao's society. so when I say 'we', I mean the Chinese population of Macao as well. If the society can hold onto its separate identity, its way of life, if it can stay a place of free thinking and free expression, a place with a history of Western and Eastern experience accumulated during so many years, with a fraternal outlook on life, I think we will have a bright future. Macao will remain a happy place for our children to grow, develop and settle.

Macao can't compete with places that are bigger and richer than us. We can't fight Hong Kong, one of the most important financial centres in the world. But why would we want to? Guangdong is a rich place now, too. In the Pearl River Delta, Macao can play a very important role, in education and in culture. Macao hasn't achieved a high level of education yet but that's partly because after the college of St Paul burned down, we didn't have a university until the 1970s. Now we have the University of Macau, which offers a range of good courses, from science to humanities, plus the Polytechnic Institute, the Open University of Macau, and so on. We've really got the opportunity to develop a local intelligentsia with room for them to develop their skills. I think that will be very important. We have to make the best attempt we can to increase tertiary education in Macao, because it's an important part of Macao's past.

Eventually it's knowledge we can sell to other countries, too, especially in the region. Just today, I read a report about our university. Many of their courses are now rated highly in terms of international standards, and students from Europe and from all parts of Asia are coming here to study—a lot of Africans, students from Sweden, Norway, and other parts of Europe, as well as from all over Asia. It's a good beginning. Macao has a long tradition of learning. We don't have the industry, we don't have the agriculture, but we do have the knowledge.

Photograph courtesy of the Historical Archives of Macau

Mrs Chen

Mainland New Arrival

Macao's population has swelled and declined through its recent history, largely in response to events in mainland China. At times of political turmoil or economic change, mainlanders have flooded across the border into the enclave to seek refuge or to find new opportunities in Macao, a transparency illustrated by the fact that almost half of Macao's present population was born in mainland China.

China's economic reforms, and gradual opening up to the outside world in the late 1970s and 1980s, encouraged many people, especially those with relatives across the border, to try to come to Macao. Leaving China officially could be a long and frustrating process of uncertain waiting, all the more difficult to accept when it involved the separation of husband and wife, or children from parents.

Mrs Chen's experiences are fairly typical. She was born in a small village in China's southern Guangdong province, which shares a border with Macao and Hong Kong. She was always determined to escape the village and live outside China, and she finally seized her chance by marrying a man from Macao. Actually joining him there was no easy task—in the end Mrs Chen and her new husband resorted to having her crossing the border illegally.

The early days of her new life were tough ones. Without proper papers, Mrs Chen became one of the growing number of illegal immigrants from the Mainland hiding in Macao's poorer districts and struggling on the fringes of the legitimate workforce to make ends meet. Her fortunes only changed when she joined a mass gathering of illegal immigrants in Macao on 29 March 1990. Police say more than 37,000 people formed an illegal gathering. Finally, they were issued temporary residency. More than 33,000 of them, including Mrs Chen, subsequently received full residency rights in Macao. Mrs Chen has now settled in Macao with her husband and their two children, part of the new face of Macao society.

I WAS THE THIRD IN A FAMILY of four children; two older brothers, then me, then a younger brother. I studied until I graduated from secondary school, then started work.

·I was a child in the 1970s. There were no buildings more than two-storeys high in my village. We lived in a one-storey building. There was no electricity until the 1980s. My parents had to go to work early in the mornings so we four children looked after ourselves. We got ready for school on our own, and afterwards we did our homework by ourselves too. There weren't any recreational things at all for us to do, just school work.

In the 1970s, life for children in China was very difficult. We only ate two meals a day. My mother worked in a garment factory, operating a sewing machine. My father worked in a restaurant as a cook. Later on, he got a job in a sweets factory. In the summer and winter holidays from school, I had a temporary job in the same factory as my father, when I was only ten years old. I had to wrap the sweets by hand. I just sat there and wrapped sweets all day. It wasn't difficult to do but it took a lot of patience. I received a few cents a day. At that time, there was no television, nothing to do after work. My only dream was to get out, to find somewhere better to live.

When I was a teenager, there was no variety at all in our daily life. The only places we heard of—and people talked about them as if they were Heaven—were Hong Kong and Macao. I knew there were other places beyond them, but they seemed completely out of reach, impossible dreams. As a teenager, I set my heart on getting to either Hong Kong or Macao.

It was our neighbour who made it all come true. She knew me, and knew I was a decent girl. She introduced me to her nephew, my future husband. I had other boyfriends, but I was so determined to leave the village, I wouldn't marry anyone local. I waited, and as soon as I heard about this relative of my neighbour's from Macao, I accepted her offer to meet him. He lived in Macao, and because I had dreamed of getting to Macao for so long, I was interested in him at once. I wanted to marry him as a way of getting out of China.

My husband's father and mother were originally from Hong Kong, but they had lived in China for a while, and gave birth to my husband there. He was still very young when they left China again and went to live in Macao—but technically, he was a migrant too. He had come to visit our village with the aim of finding a wife. His parents told him the girls in Macao and Hong Kong might not be faithful, but the girls in China were more innocent and made better wives. That's why he came looking in our village. Personally, I think you can find good girls everywhere, but a lot of people think the same way as my husband's parents.

I was twenty-six when I was introduced to my husband and he was the same age. Because of the problems with permits, I couldn't leave the village, and I couldn't leave China to see Macao before marrying. So I waited in the village for a year after we met, and then married him. It was blind faith. He came to visit me in China once a month during that year before we got married. But I still barely knew him. My decision to marry him was based on pure intuition. I trusted him.

Right after the wedding, I still couldn't apply to come to Macao. I gave birth to my first child in China, and when I was twenty-nine, when my son was only six months old, we decided we couldn't wait any longer and I should go and join my husband in Macao. I wasn't sure what our new life would be like so I left the baby behind with my parents while I went to find out. My parents were really worried but they said: 'You make up your mind and make your own decision. But if you go, don't take the child with you—leave him with us for the time being.' So I did.

I arrived in Macao as an illegal immigrant. I was very frightened. The problem was that China restricted the number of people who could leave, and if I'd done it legally, I would have had to wait years and years to join my husband. The Macao government welcomed me—there was no problem on their side. We had to bribe someone to let me cross the border. I just walked over the border, I didn't swim. It still happens now, people doing exactly as I did, although not as many, because now China allows 300 people every day to go to Macao legally, so people aren't as desperate to take the risk.

Along the side of the border, there is a little path, and people can just walk through. If you pay the right people, the guards will turn a blind eye and let you cross. When I did it, it was scary. I had to travel at night. My husband knew some people in Macao who helped us, and they told us what time I should go. My husband crossed the border into China first and took me with him to introduce me to the guards who were going to let me pass. He came into Macao legally with his residence papers, through the immigration gate. Then I walked down this small illegal path on my own, past the guards, and joined him.

When I walked over that border and saw Macao, I thought I was hallucinating. The reality wasn't much like the dream. Now, Macao has developed and it is a great place, but then, it wasn't much. My dream had been of a city full of beautiful skyscrapers. In those days, there was only one area with three- or four-storey buildings, and we didn't live there. We lived in the slums.

Our home was a squatter hut—they have all been knocked down now—cramped and overcrowded. We had one room of our own, partitioned off from the others inside. A bed, a dining table, a backless stool, and that was it, the space was already full. We had to cook and do everything inside the room, and go out to use a public toilet. That first night, I was so sad. All my dreams, and this was the reality! I wasn't happy at all. But I decided I would work hard and make enough money to get us out of there.

My husband had told me we would live in a wooden hut, so when I was still in China I had looked through magazines and had seen pictures of the sort of wooden houses they have in Japan. I thought he meant a big house like that, a chalet. This wasn't at all what I'd expected. This was just like a rubbish tip. It was all partitioned—each family rented one section. Some of the young people went out to work, and the older people stayed around the house all day. I couldn't go to work because I had no papers, I was illegal. There were lots of women like me— illegal, but married to husbands here. We took badly paid work to do at home.

Garment factories in Macao at that time used to employ people to cut off all the loose threads by hand to finish the garments. My neighbour got this kind of piece-work and brought bundles to my home too, and I did that, trimming and cutting. I worked every day from 7am to 2am the following morning. All I could think about was making money and getting out of that place. I had to sit all day on that backless stool—the pain in my back! And in my fingers too. My husband and the neighbours all went out to work—my husband was a labourer on a construction site—and I sat there on my own all day, cutting and cutting. Some old ladies also came and worked with me sometimes.

I was lucky in some ways. At least my husband was honest and wasn't married to someone else when he met me. I saw many cases of young women who were cheated by men who were already married, but came to China pretending they were single and looking for a wife, and went through a false wedding ceremony. There were so many of us young women eager to get out of China, it wasn't hard to convince us. The women left China illegally, like I did, thinking they were joining their husbands, and found themselves stuck in Macao—with no work, no papers, and no way of getting back to China. Sometimes they already had babies or were pregnant. It happened to some of my friends here. They gave the babies to the nuns to look after, and either stayed here on their own or found a way of sneaking back to China illegally. There were lots of cases like this.

Many people crowded into the slums, many of us had come from China recently and couldn't go out to work because we were illegal. Soon I started to make friends with other young women in the same situation, and we shared our problems. The police didn't bother us. They came in from time to time, and they knew who we were, but as long as we stayed in that slum area and didn't try to leave, they were very kind, they didn't cause trouble for us. I was very worried about my son, left behind in China, and very worried about myself too. I couldn't see any future for us. I just couldn't see how I was ever going to improve things.

Then everything changed. Big news. In 1990, I heard the government was offering to register illegal immigrants and on a certain day, from 9am to noon, any illegal immigrants who declared themselves to the government would be given legal papers. That day was tremendous! There was a big rally in Macao—everyone came out of the slums and crowded there to get their identity papers. That night, I had been asleep at home when some people came to the house and told me about the amnesty the following day. I got up at three o'clock in the morning to go and queue up—so early, but the streets were already crowded with people, there was such a long queue. We all went to the greyhound racing stadium and gathered there. We filled the whole place. Everyone there got their ID card from the government. The news had even got through to China and people were coming across the border illegally that night to join the crowd and get their papers too.

Once I had those identity papers, I could get a proper job in a garment factory. It wasn't much money but a lot more than I was making before. We bought new furniture—proper chairs instead of that old backless stool. We moved out of the partition room and got a new place, also a wooden hut, but bigger and not shared with other people. I had been in Macao three years by then. It had been a really tough time. I worked full time in the factory for more than three years, and had a second job in the evening, working in a school as a cleaner. We had no savings. We were trying to build a home. As soon as we got paid, we would go and buy something else for the house, little by little. Finally, I got a job in a fast-food restaurant that had better pay than the factory.

I'd had a second child, a girl, the year after I came to Macao but I hadn't been able to look after her myself. Because she'd been born in Macao, she was legal here. I sent her back to China for my mother to look after. She spent her first three years with my parents and her older brother. I missed them terribly. Whenever I had a little spare money, I asked people to take it back to my mother in China for things for the children.

The year after I got my papers, I brought my little girl back to Macao to live with us. She could come because she had Macao residency, but my son had to stay behind in China and wait much longer. He was only able to come and join us—finally—last year, when he was already eleven years old. When my daughter came, I had to leave my job at the fast-food place to take a part-time job in the factory, so I could look after her. Things have improved steadily since then.

It's been hard. Never enough money, always difficult to save. The place we live in now is better than before, but we have to pay a lot. We're in a government housing unit. I'm happier than I was before, but I feel a heavier sense of responsibility in some ways. I worry about the children's education and finding enough money for them.

Most people in Macao have really loving hearts. They accepted us from the beginning, from the day I arrived. The officials in the Portuguese government were very helpful and kind. It didn't feel strange at all, the fact they were Europeans. I just hope nothing will change now, after the reunification with China. The system, the way everything's organized, I hope it all stays just the way it is. My wish for my son is for him to be a dragon—that's what we say in Chinese—it means I want him to be successful. I don't mind where the children settle down or what they do, as long as they have a good education and make a useful contribution to society.

When I think of my life here and my family's life in China, it's hard to explain what the biggest differences are. Education is one thing. It's all free here—in China we have to pay. And the public services here are better than in China. The Macao government takes better care of poor people. If I'm sick, the government lets me have free treatment. When we couldn't afford to rent a home in the private sector, the government gave us low cost housing with a low rent. I've never seen that happen in China. They even help with the cost of education. Although schooling is free in Macao, we still have to pay for textbooks, exercise books, and uniforms. But if we haven't got enough money, we can tell the school and they help out, either they grant a complete exemption or we can delay payment until later. That never happens in China. In China you just have to find the money, they don't care how you find it, you've just got to. When I was in school, my mother had to borrow money from all over the place to keep me there. But here, it's different.

The government in Macao has really cared for the refugees, immigrants from China, like me. I know it's got to keep a balance. If too many other people come, the society won't be able to cope. So it's got to be regulated, it's got to be balanced.

My three brothers are all still in China. The oldest one manages a restaurant, the younger one owns his own restaurant business. We've all had hard lives—I've had some really tough times coming to Macao— but, all in all, I think all the suffering I've been through, all the dramas, all the changes, have been worth it. I've got my two children with me now and they've got the right to grow up in Macao. That makes it worth all the suffering.

Chen Gongan

The New Generation

Gongan is the son of Mrs Chen. He was born in China, but has now
settled into life in Macao as a newly confirmed resident of the enclave.
He came to Macao in 1998, at age eleven, when he finally received
permission from the Chinese authorities to join his parents and little sister
there. He spent his early childhood with his grandparents in their home
village near Guangzhou.

Gongan is part of Macao's next generation. By the time he finishes
school, Macao will be a Special Administrative Region of China. His
identity now is strongly Chinese—with minimal awareness of the
Portuguese 'foreigners' in charge when he arrived.

THE VERY FIRST THING I remember is being in kindergarten in China.
I was about four years old and I had to go there all day. We used to do
handicrafts, cutting and folding paper, and learn how to count, and
how to write simple Chinese characters. We had two hours of playtime
every day, and at lunchtime we all had a midday nap. My grandmother
used to come and fetch me from school—but when I turned six, I started
walking to and from school on my own.

I lived with my grandparents and my two uncles. I was the only
child at home. There weren't any children nearby to play with. Some
other children lived in the same building but everyone kept to
themselves, they didn't play with me. In the evenings, I watched
television. I liked the cartoons best. They put those on from five in the
afternoon, after school.

I knew my mother lived in Macao and I used to think about her a lot,
although I couldn't really remember what she looked like. We had a
photograph and I used to gaze at that. I was always dreaming about
leaving home and setting out to Macao to look for her. I used to beg my
grandmother to let me go but she always said the same thing: I couldn't.

Then last year she sat me down and said the time had come, I could go. I was so happy.

My first impressions of Macao were that it was very beautiful. Guangzhou isn't as pretty—it's very noisy and overcrowded. My mother came to China to collect me and we travelled to the border by bus. Since my mother got her own papers a few years ago, she'd been able to come back to China to visit us so I got to know her again a little bit.

Macao is much more fun than China. My friends at school are better friends than the children I went to school with in China. We go out and play and do things together. The children in China never did that. And the food's much better. I like meat—in China, there was lots of rice and not much meat and vegetables.

The school's better, too. Here they have air conditioning and lots of lights. In China we had to keep the windows open because there was no air conditioning, just one fan, and only two tube lights in the ceiling. The size of the class is different too. In my Macao school, we have about forty students in each class. In China, there were usually about sixty of us learning together.

The style of learning is much the same—but the teachers here are much kinder than in China. In my Chinese school, they were mean, very strict. They used to hit us on the hands with rulers, but that's not allowed in Macao. Here the teachers are more patient. They give us a special book to show our parents, so my mother can check my homework and the teachers can make remarks. In China, she never knew what I was studying.

I had just started to learn English when I left China—I'd only done two weeks. So I was really behind in English when I started in the new school. I had to work really hard and I've finally caught up. But I was better than the children here at writing Chinese characters, and good at mathematics too. I like gymnastics best. I started doing it in China but we only had a one-hour class a week. Here it's two hours a week. I like basketball, which I'd never played before. After school, I play table tennis on my own against the wall, watch TV, or go out sometimes with my new friends.

I play with my little sister now, too. I remember when I found out she was going to live in Macao with Mum, long before I could go. I felt very strange about it. How come she could go and not me? She came back to China occasionally with Mum or Dad to visit us, but she never said much about Macao, and what it was like, and I didn't like to ask questions.

I want to be a police officer when I'm older, here in Macao. They're great. I like exercising and travelling round different parts of Macao, and as a police officer, you can do that.

I know all about Macao going back to China, I heard them talking about it on television. I'm very excited about it—we're going to be reunited with the motherland. I know Macao has been under Portuguese administration but I don't know anyone Portuguese, I don't think I've ever met one of them. I think some of the police officers are Westerners but most of the ones I see are Chinese.

Sometimes I miss China, I don't know what it is, but although I miss it, if you'd ask me how I'd describe myself now I'd say: I'm a Macao person.

Jorge Smith

A Taste of Macanese Cuisine

J orge Smith, now in his mid-thirties, was born in Mozambique to a Chinese mother and half-Chinese, half-Zimbabwean father—but he has links to Macao through his grandmother, who grew up on Macao's Taipa Island. He spent part of his childhood in Lisbon, escaping the political upheaval in Mozambique, and went on to study hotel management.

He came to live in Macao in 1992, to take part in the refurbishment and reopening of Macao's most famous hotel, the Bela Vista, a grand old colonial mansion hotel. In the past century, it has taken on a number of different roles, from refugee centre to hospital. It is one of Macao's most famous landmarks, and after the handover becomes the residence of the Portuguese Consul General in Macao.

Jorge Smith is now the food and beverage manager at the Mandarin Oriental, Macao, where he has developed a passion for Macanese cuisine, a cuisine developed through the centuries in Macao, and which many see as an important feature of the cultural mix the Macanese community has come to represent, but which is now under threat.

THERE AREN'T MANY PLACES in the world like Bela Vista. Physically, it's an impressive building. Not in the same way as the world's big hotels, the Peninsula and the Grand Hyatt and so on. It isn't gigantic, but it is very old world and it's done in the way it must have been when it first opened. The restorers did a lot of homework. It's an old, colonial mansion, serene, tranquil, and hugely romantic.

The Macao suite has the best bathroom in the world. You walk in and want to run a bath and lie in it for three hours. It's a free-standing bath in the middle of the room, not off to the side, standing on four legs, with a central drainage hole. It's just a fantastic place. And the suite's got this very old Chinese reclining chair in it. Words aren't good enough to describe what the Bela Vista is like.

Pretty soon it will be the home of the Portuguese Consul General, and there are mixed feelings about that. I think it's a shame they're closing it to the general public. I know the amount of money invested in the place, and for it not to be open to the public is a shame. If you ask people born in Macao, they'll say it's a very important building. It's been everything from a school to a hospital—and between times, a hotel! But on the positive side, it is a symbol of Portuguese presence—the fact it was built by English people is irrelevant! Much of the money spent on it was paid for by the Portuguese administration, so maybe they have the right to say: 'It's ours. We spent money on it. It's a small jewel in our crown.'

If you ask me if they made the right decision or not, I'd say no; partly because the upkeep of that building is terribly expensive so, from a practical point of view, I'm not sure it's reasonable. From an image point of view, no doubt it's great. But they could have taken this opportunity to develop any one of five hundred other beautiful buildings in Macao, including the old Chinese buildings. They're renovating all these beautiful Western colonial buildings, but at the same time, the beautiful old Chinese buildings are deteriorating.

Until I was nine, my first language was Cantonese. My great-grandfather was one of the first Chinese people to settle in Mozambique. He was originally from Guangdong province, and first he came to Macao, and settled here. He was a cook on a ship and travelled, ending up in Mozambique. I empathize very much with the Macanese people, because the same sort of thing happened to me. I'm Chinese by blood, but I was born in Africa in a Portuguese colony. I started learning Portuguese as soon as I went to school. It was Chinese in the mornings, Portuguese in the afternoons. Then it was Portuguese all the time when I went to Portugal; then English when I was sent to school in England. So what am I? I really don't know. What is my spirit? I'd say a bit of everything. I'd compare myself to a Macanese in that way.

I was shocked when I first came to Macao because it was so Chinese. I thought of it as more Portuguese but, of course, this is China under Portuguese administration, and only a tiny fraction of the population even speak Portuguese. It's so different from Hong Kong, which was a colony. If you go down a road, there's nowhere else in the world that has this architecture, this mix. It's a rich blend of East and West, the way the Macanese people themselves are both Eastern and Western.

The Macanese have had their own identity, culture, practices, and language, but ask most ordinary people in Macao and they'll say that anything related to the Portuguese isn't good. Just yesterday I was catching a taxi and chatting to the taxi driver, and he said: 'Look, the Portuguese here, they're just disrupting our way of life.' They can't wait for the Portuguese to leave so Macao can keep on growing. The average person's view is: what have the Portuguese ever done for us? The British left Hong Kong with a tremendous apparatus of government. There was a great feeling of apprehension about the British leaving Hong Kong. But in Macao, there's relief, rejoicing that the Portuguese are leaving.

We have to look at Portugal in terms of a modern country, and in those terms, it's only some twenty years old. In the past twenty years or so, they've been trying to do what they hadn't done in the past four hundred. I don't blame the Portuguese administration for what they're doing now, but it's too late and too little. At least they're trying. I think during these past four hundred years, they left some fabulous buildings in Macao, and some great food, but is that enough to nurture a society? I don't think so. The Portuguese have now realized they didn't do enough in the past and they're trying hard to make amends. Macao right now is in a real stew—there's a certain tension. The Portuguese are trying to leave some sort of legacy behind—but I don't think it will last. The Portuguese influence never really was all that pervasive in the first place.

In the short term, I'm very concerned about the rule of law. I'm in the tourism business. Every time a bomb goes off, imagine a thousand fewer people coming to Macao. In the last couple of years, it's been horrific, with all the casino-related people jockeying for position. The feeling is that once Macao is handed over, the Chinese administration will sort things out and take no prisoners. There's a lot of optimism for the future. That's very important—more important than whether or not Portuguese culture will be preserved. After all, most people in Macao are Chinese.

From a business point of view, from a marketing perspective, Macao must play to its European feel. It's a niche position they have to maintain, which differentiates it from any other Asian city. I think the Chinese recognize this. Look at the debate in Hong Kong about maintaining the culture. Here, we already have that sense of local culture—open your eyes and you can see it.

In the last few years, one of the fashionable terms to use about food is fusion cuisine, referring to a mixture of ingredients from different places coming together to create something new. Macanese cuisine is the original fusion cuisine. It grew from the history of Macao. You had Portuguese people coming here over the centuries who wanted to eat something similar to what they had back home and they used local ingredients to try to create that. Throughout the years, with intermarriage, the Macanese community evolved, and the food evolved with it. This original Macanese cuisine is heavy, home cooking; tasty but nothing fancy, and quite rustic, not terribly sophisticated.

When I was in Mozambique, one of the dishes I really liked was the grilled African *piri piri* chicken—marinated chicken, grilled, with chilli oil brushed on it. In Macao, you have a local version of African chicken. People like to say it reflects the Portuguese maritime journey to Macao. The sailors turned the Cape of Good Hope, sailed along the African coast, via India, where they picked up all these spices that they brought to Macao. Here they created this spicy tropical marinade for flavouring the chicken before it is grilled.

My favourite dish is *tacho*, the Macanese hot pot—a little bit like a Portuguese hot pot, *cozido*, but with Chinese sausage substituted for Portuguese sausage, and with pig's feet. It has a Chinese feel. The idea is the same but the flavour is completely different. When we talk about European cuisine, a great deal of it came from Arab cuisine because of the Moorish influence, which really shaped southern Portuguese cuisine as we know it today. The Chinese sausage, like the Portuguese sausage, is made from pork and blood, but it has a completely different flavour. So the final result is something completely different—a Portuguese person wouldn't recognize it—including cabbages and tripe, which are not in the Portuguese version at all.

A celebrated feature of Macanese cuisine is the *cha gordo*, or Macanese high tea. You can translate it as 'rich', or sumptuous, tea. I remember having those types of teas in Mozambique. My grandmother used to invite people over in the afternoons, for something that was not lunch or dinner but something in-between. It was very special. We'd eat everything—fried noodles, roasted pigs, fish—an enormous buffet of things. We just helped ourselves to whatever we wanted at our own pace. And a lot of desserts, cakes with cream, fruits, sesame fried balls— all the good stuff set out on one long table.

I didn't realize this was in itself a special type of meal. When I first came to Macao, people kept telling me about the *cha gordo*, and finally I realized this was what they were talking about. It's a family affair; uncles, aunties, and all the cousins—imagine twenty or thirty people talking and eating away. It doesn't exist in Chinese culture, or in Portugal. It's something exclusively Macanese.

Traditional Macanese people also have special decorations. Some people would call the decorations kitsch—they're extremely ornate, very rich. It's their idea of what luxury is about. Imagine the plate, with a Chinese motif, but drawn by Western hands—then you have something that is Macanese. Doilies, very ornate, embroidered tablecloths, colourful teacups, and afternoon tea services.

I once went to a *cha gordo*, invited by a good Macanese friend of mine. It was a feast. We started at six o'clock in the evening and I left his house at around one in the morning. It's neither lunch nor dinner— you keep eating on and off for five or six hours. The importance placed on food is extremely Chinese in its orientation—he said, 'You should never save on food'—but the format was very European, suited to the convivial aspect of Macanese society.

On this occasion, there were about twenty dishes set out. We had turnip cake, which is boiled turnips, grated, and marinated in all sorts of things. Then Chinese sausage, diced and mixed together with the turnip; you add chives and bake the whole thing with soy sauce on top. We had chicken rice—chicken with barbecued Chinese meats, sausages, rice, and tomato—like a dry risotto, with all the ingredients put into a pot and cooked the way the Chinese would cook it, with a lid on for about half an hour until everything's done. You don't need chicken stock because the chicken's right in there, the flavour goes into the rice and the rice is cooked together. The Portuguese have the tomato and the rice and chicken part—the Chinese add the sausage and the cooking method.

We also had some *dim sum* for want of a better phrase, including tasty *abebicos*, a *dim sum* in the classical Cantonese tradition, with the same sort of skin and fillings found in local *dim sum*, but distinctly Macanese in that the skin is folded into a different shape. We also had samosas, which wouldn't occur to a Chinese person to include in the spread. Afterwards, we had mango pudding, home made by my friend's auntie, very rich and sumptuous. The idea of the pudding is Western but the mango, the tropical fruit, reflects where we are. They even set out a traditional embroidered tablecloth that belonged to his auntie, edged with handmade lace. It's very specialized work, and very Macanese. About twenty people gathered together, aunties and cousins, and friends, to celebrate the father's return from Portugal. A typical reason for a *cha gordo*.

That's the old-style Macanese cuisine, and it's not eaten as much nowadays. People didn't write down the old recipes, and they are already forgetting how to make these dishes. Part of the reason is that many people in the old Macanese community have moved on. The Macanese have always travelled, but lately, in the run-up to the handover, the exodus has been significant. They leave—and these recipes go with them. There are maybe two or three restaurants left where you can really get that traditional Macanese cuisine—and, of course, in Macanese people's homes.

What we're seeing instead is a new form of Macanese cuisine. There are Portuguese restaurants where the food is cooked by Chinese people. A good example of that modern Macanese cuisine is the egg tart, a popular treat recently, based on a Portuguese recipe baked the Chinese way. If you go to eat egg tarts in Portugal, they're extremely rich. Here, they are much lighter, and not half as sweet; that suits the local palate. Now, the egg tart is becoming famous, it's already doing well in Hong Kong and Taiwan.

The original Macanese cuisine evolved gradually over several hundred years. The modern Macanese cuisine is a Chinese person's interpretation of what a Portuguese recipe should be. That's an entirely different thing.

Sin Dohan

A Fisherman's Story

Mr Sin, born in 1934, is passionately looking forward to Macao's return to China. He has spent most of his life as one of Macao's fishermen, living and working on board the family boat with his wife and children, and only coming close to land intermittently to sell their fish or seek shelter from a typhoon. He applauds the Chinese Communist Party for supporting the industry and protecting their livelihoods.

He and his wife are typical of many fishing families who have finally made the decision to live on land, the first to do so in very many generations, as a result of the decline of the fishing industry in Macao and changing fishing habits. The transition has been difficult. Both Mr Sin and his wife are illiterate, and life at sea was, until recently, their only experience. Nevertheless, both are delighted now to have made the change—and pleased, too, that their children have all found work in other industries.

I WAS BORN ON A BOAT, off Taipa Island. As far back as anyone knows, for centuries, I should think, everyone in my family fished and lived on their boats. I'm the very first generation to leave fishing and go to live on land, but I worked on a boat, making my living from the sea, for decades before that. I started a long time ago, even before the Communists took over China.

Our family boat was a small fishing junk without a motor. When I was a child, three generations lived together on the boat—my two grandparents, my parents, me, and my brothers. We lived at sea, in the boat, all the time. When the weather was good, we'd be out there fishing. If a typhoon blew up, we'd come close to shore and take refuge in a typhoon shelter. The typhoon shelter we used is where the racecourse is now. The boat wasn't big—maybe 18 feet in length. We were very poor. Inside there was a simple partition, just a piece of board, making two rooms. No privacy on a boat. The unmarried ones, us children, slept with our father and mother.

I never went to school. I can't read or write. The only thing I can read is my own name. Our children didn't go to school either. We had four sons and they went to primary school while we were in the typhoon shelter. When the weather would get better we would sail off again, and they came with us, so no more school.

My childhood memories are all of fishing. We'd go out to sea, cast the nets, wait, and then haul them in and collect the fish. It was very hard, physical work. We didn't have motors like they do nowadays, everything was done manually, and the waves were often really high. The boat was small, so if a wave broke, the whole junk was soaked; the bedding, clothing, and everything got wet.

I remember the war with the Japanese. They came to Macao and all round this area, and made our lives very difficult. Those were hard times for all the fisherman. I was about seven then. During the Japanese occupation of China, we couldn't go out fishing far from shore. It was too dangerous. If we came across a Japanese patrol boat, they would have killed us straight off. Many people were killed that way in the war years, so many. If a Japanese warship came across a fishing boat, it would open fire and kill everyone. The fishing channels were all mined as well. Our boat was very small so we could sail over the mined areas safely, but the bigger boats were in real danger. The Japanese had speedboats, too. As soon as they saw a little fishing boat like ours, they'd come straight across and ram it until it sank.

In 1949, life got better. By then, we'd survived the Japanese occupation and the Kuomintang, who took over immediately after the Japanese. It was only when the Communists came to power that our lives really started to improve. No one robbed us any more; we could go out to fish, and we were free, we could feel safe. Even under the KMT, the robbing went on. Immediately after the war, we often ran into pirates who stole whatever we had—our catch and anything else. Once the Communists took over, that became much less of a danger for us, and by about 1952, that threat had disappeared altogether. We were never bothered by pirates again.

We mostly fished for Macao sole and yellow croaker, which is used to make a well-known Cantonese dish. It's a salted fish that is very popular here in Macao. The best years were the late 1960s to the mid-1970s. There were plenty of fish then. Now there are far fewer. I think it's mostly because of the increased pollution in these waters.

We were never rich. Even in the best years, when there were plenty of fish to catch, the price of fish was low. At least I earned some money, and life was less difficult than before.

My father gave me a boat when I got married. He bought another one for my younger brother, and went to live on it with my brother's family. When my father died, I inherited another boat from him, and I passed it on to one of my own sons. I have four sons—they're all married now.

We came back to shore every three days or so to sell the catch. If we caught crabs, we might come back straightaway to sell them to a wholesaler. The other fish, the yellow croaker, we hung and then salted and stored in the boat, and took back to sell to the wholesalers a few days later.

In the beginning, we were really poor. In 1951, we joined the fishing association in mainland China. We got a licence from the Chinese authorities so we could fish around the coast of Guangdong. Each licence lasted several years at a time, then we had to renew it. That whole area was teeming with fish. Whatever we caught, we'd sell a certain portion of our catch back to the Chinese association at a very low price. The rest we could sell to anyone, so we sold it to the wholesalers back in Macao for a much better price. That was until 1984. We had to sell the Chinese association so many fish a year—they set a quota—but after 1984, they released us from that and we could do what we wanted, sell them some of our catch or not.

From the late 1960s onwards, we started to install motors in our boats. The whole process took a long time. The Macao government didn't help us financially—we borrowed from the wholesalers. It was only in 1978 that I was able to buy a motorized boat. It was 150 horsepower and cost about 200,000 patacas for the boat and the motor. That was a lot of money in those days.

There are far fewer people fishing nowadays. That's partly because this area has fewer fish, the pollution has killed them off. Fishermen have to go far away to find a decent catch, and that means they need a good-sized boat. In our day, there were lots of fish right here, close to shore. For example, the place where they've built the new Macao airport, that used to be a rich fishing ground. I remember that you could sail there and just drop your nets, and get a tremendous number of yellow croaker straightaway. Now, there aren't any fish there at all because they've built the airport.

The economic development in Macao has meant many fishermen can't carry on. They can't work close to shore any more, they have to sail so far away, so they sell their boats and move on land to work. A lot of them have work on the building sites, as construction workers, or in restaurants.

I sold my boat nine years ago and we came to live on land. I stopped work as soon as I came ashore. Every day I go along to the fishermen's association here in Macao and help out there. Sometimes I go across the water to the mainland Chinese fishing association there for meetings, representing the interests of the Macao fishermen.

Many fishermen just don't want to live that kind of life any more. They would rather live on land, watch television in the evenings, and have an easier time of it. Young people are different now anyway. My grand-daughter is at university. Life isn't the way it used to be. Not one of my sons is a fisherman. One works in a cargo company in Coloane Island, two of them are drivers, and the fourth is a foreman in a company. One of my daughters works in a museum in Macao. We're happy with that. I'm glad they're not forced to go out fishing to earn a living. They have a steady income, and they don't have to endure the wind and the rain out on the boat.

Fishing is a hard life. Before we had a motor, it was very tough. The wind changes direction often and the bigger boats, that go out far from shore to get better fishing, can easily get caught out there in a storm. They get stuck out there, too late for them to get back to the typhoon shelter.

China's economic reforms allowed both Hong Kong and Macao fishermen to quit fishing and move ashore. If China hadn't opened up like that, we wouldn't have had the chance. We'd have had to carry on living on the water for the rest of our lives. China's opening up made it possible for Hong Kong and Macao to develop economically and become more modern. Macao's building boom made it possible for my sons to go to land and find jobs as workers on the construction sites. My sons could earn twice as much on the sites as I could earn fishing!

When I was still a fishermen, we never lived on the land, just on the boat, that was our home. Since we came to live ashore, I have bought four flats. My wife and I live in one with two of our grandchildren. My sons and their families live in the other three. They're paying me back for the flats by instalments.

We bought the first flat in 1983, and another one in 1988, when I was still fishing. Then, when I moved ashore for good, we bought two more. We used to have a bank account and keep our money there. We couldn't keep it on the boat. It wasn't safe, there was always the danger of pirates.

I remember the strong typhoon here in Macao in 1983 that destroyed a lot of fishing junks. At that time, we had a little boat. The typhoon really scared us, and that's when we made our first plans to move ashore and start living in an apartment instead. There were many accidents at sea. At night, there were collisions between the boats. Recently, two fishing junks rammed into each other and one man died. One of the boats sank. Even the big tugboats collide sometimes. It's nobody's fault, just a mistake. They can't always see each other. If a big ship hit a little one like ours, we'd sink for sure.

Fishermen are delighted about the return to the motherland. The Portuguese have been here for more than four hundred years already. All the residents of Macao, including the fishermen, hope the handover will be very smooth. The Portuguese did help the fishermen in the last ten years, they were very kind; but they didn't do much before that. Now the officials at the Department of Marine Affairs come to our fishing associations to visit us, and sometimes invite us to go to their offices to let them know if we have any problems. We're putting up ten candidates, representing the fishing community, as potential delegates to the handover committee. Ten fishermen have already put their names forward for selection.

Since Deng Xiaoping's day, there's been a lot of dialogue between the Portuguese government and the Chinese, and we fishermen have been involved too. We sent our representatives to Guangdong province to take part in some of the political meetings about the reunification, meetings at a provincial level.

The handover takes place on 20 December and I'm counting every day. The fishermen are already making plans for something special for the day of the reunification. We're planning a sort of procession through the harbour of all our boats, the same sort of thing as they did in Hong Kong in 1997.

I can tell you exactly how many days there are to go until 20 December! I've got an abacus at home, counting the days off, backwards. Every day I go to it and have a look: one day less to go, one day less! Every morning I move it on a day—one day less!

Fong Gamho

Fisherwoman at Sea

Mrs Fong Gamho, was born in 1934 on board her family's junk. Like her husband, Mr Sin, she has seen enormous changes to fishing practices in her lifetime. The closer relationship that gradually developed between the floating fishing community and the administration systems on land in the latter half of the twentieth century, led to her giving birth to her children—four sons and two daughters—in clinics and hospitals in Macao, rather than on the boat.

Ultimately the changes allowed the family to move on land after generations at sea. Mrs Fong is clearly delighted with the transition; life at sea was filled with physical hardship. Girls and women worked alongside the men, contributing to the fishing as well as fulfilling domestic tasks on board, such as making the family's clothing, cleaning, and preparing food.

Some analysts suggest that women in these traditional fishing communities often enjoyed a higher status and a greater say in decision making than women on land, precisely because they made this direct contribution to the family's livelihood—they were thus valued more highly. Certainly the marriage bond at sea must have been critical. A girl stepped from her family's boat to that of her husband's family when she married, after a community ceremony, and had to adapt quickly to her new role, with a new family, in the confines of a small boat. The isolation of life at sea made contact outside the immediate family unit, and the support of a female network of relationships outside her husband's own family, difficult.

I WAS BORN ON A JUNK in this area of southern Macao, in a typhoon shelter. It's gone now—they reclaimed the land about twenty years ago. Women have to do more work than men, on the boats. I started when I was a child. My elder brother would cast the nets and haul them in and then it was my job, because I was a girl, to collect all the fish. They were big fish sometimes. I picked them out of the nets, one by one, and gathered them all.

I don't remember a lot about the war but I do remember the Japanese war planes flying overhead; everyone was frightened to death. We rushed to hide at the bottom of the boat, underneath our belongings. We were so scared!

I married my husband when I was twenty. He was twenty-two. He came to get me with his fishing boat. It was a long sail to where we were moored, waiting. He sailed right down the coast from mainland China, to take me on board with him. That was our wedding! That day he arrived to marry me was the first time I'd ever seen him. It was an arranged marriage. That was the way things were done in those days.

The matchmaker was his cousin, and she was married to my cousin. She arranged the match with our parents. I don't remember feeling anything in particular that first time I saw him, after all, marriage is marriage. It was just like meeting anyone for the first time. You see someone, realize who he is, and you both just say hello to each other. It was just like that with him. That's him, he's my husband now, let's get on with it!

I'd already met his parents when I joined his family. They came to visit my family's boat with the matchmaker, to have a look at me and talk to my parents about arranging the wedding. They were very kind to me.

It's usual during a fisherman's wedding to line up all the boats of the relatives of the bride, all in a row side by side. Then the groom comes in his own boat and picks up the bride and takes her away on his boat with him. In my case, my husband was coming such a long way to get me! I had a few dozen boats lined up on my wedding day for my send-off, some belonging to friends as well as my relatives. We had a reception afterwards as well. There's a floating restaurant that all the fishing families like to go to and we all went there for the wedding ceremony and then a wedding banquet. I didn't have my own pair of shoes until I was thirty—we never wore shoes on the boat—but I wore shoes on my wedding day, for the first time ever! And so did my husband.

My father was very poor, partly because my mother was ill all the time. It was usual for the groom's family to put up a dowry for the bride, and my husband's family paid my parents 1,000 patacas for me.

It doesn't sound very much but in those days you could fish all day and only earn between 10 and 20 patacas—so 1,000 patacas for a wedding dowry was a lot of money. Life was hard.

Although I was born on the family boat, I didn't give birth to my children on our boat. I came ashore to a clinic, or the Chinese hospital here in Macao, for the births. All six children were born in Macao and they all have Portuguese passports.

It isn't easy coping with young children on a working boat. When they were very little, we tied them onto the boat with a rope harness for their own safety. When they got a bit bigger, we'd tell them to stay lying down. Otherwise they could get hurt with everything going on around them. Then, from the age of about four, they started working with us, helping out with the catch. We might still tie them up to something on the boat for a while so they wouldn't get swept overboard.

We had no entertainment on the boat at all, there was no such thing as fun. The only thing I can think of is that we sometimes had weaving or knitting competitions. Two boats would come alongside and we women would compare our handiwork to see who'd knitted the best sweater. That was it. There was just no time for amusing ourselves. In the evenings, when it was too late to fish, we'd mend the nets. It was easy for them to get torn during the day's fishing. We had a little kerosene lamp to work by.

I used to make all our clothes as well. We wore tunics with lots of pockets, that was the style. The trousers were very baggy, with no belts, so we had to tie them in a knot at the side to keep them up. You wouldn't catch any fishermen wearing clothes like that nowadays! I can do all sorts of things, although I can't read or write.

Now that we've moved to live on the shore, life is much more comfortable. Being a wife on land is much better than being the wife of a fisherman. A land wife, even if she doesn't have much money, can still find interesting things to do. If she's bored she can go for a walk somewhere, or go to a teahouse to meet friends. In all those years on the boat, I never went to a teahouse! As a fisherman's wife, the only place you can go for a walk is round and round a tiny boat.

From the moment we left the typhoon shelter and went out to sea, we were wet all the time. We hardly ever managed to get dry. Even when the sea was quite calm, the boat was always rocking about. I always had to keep my rice bowl in my two hands or it slid off somewhere. The only way to keep stable was by standing upright all the time. If you sat down on something, you'd end up sliding across the boat as soon as the next wave came. Even if you had money, it wasn't much use to you on board a boat.

I did all the cooking for the family. I had to keep both hands on the stove when it was lit and hold down everything else as well or things would all spill. We had a small wood burning stove, and carried enough split wood for about ten day's cooking.

Our diet was fish. That was it. No chicken, or pork, or any other kind of meat. We didn't have any fresh vegetables either. So we ate fish with rice, sometimes with a kind of pickled vegetable, and I cooked with oil and sugar. That was the lot. No milk, nothing. Sometimes I'd make congee [rice porridge]. Other times, we bought some beans and I cooked them with sugar and rice to make a sort of sweet congee. If there wasn't any sugar, we'd put fish in it instead and make it a salty congee.

If the weather was bad and the boat was pitching and tossing, I couldn't cook at all. It was too dangerous. So either we'd have to go without food all day or we'd have to head off to somewhere calmer, like the nearest typhoon shelter, and I'd start cooking when we got there. We didn't get ill very often. In fact, I think we get ill more now, since we came to live on the land!

The only time I miss being on the boat is when I go to the market to buy a fish for dinner and I see the price! Maybe 20 patacas for one fish! I think to myself: before, I could have bought a whole barrel of fish for that price. In the old days we could get as much fish as we wanted, just like that, free of charge. Now I have to go to the market like everyone else and pay money for it—a lot of money!

I'm much happier since we sold the boat and came to live on dry land. For a start, we don't have to put up with wind and waves any more. Life is much more secure. We're not always on the lookout for pirates or people who might crash into our boat! There's plenty of work, we've got plenty to eat. All our sons are working on land. The land people accept us pretty well, no problems there, we all get along together. Of our four sons, two have married girls from the fishing community, from families who also used to live on boats, and the other two have married land girls, from mainland China.

City of God

Religion has been central to Macao since its earliest days. When the Portuguese merchants first settled there more than four centuries ago, missionaries and priests travelled with them, eager to spread Christianity to Portugal's new trading partners. Macao soon became an important foothold for Christianity in the region, and a centre of intellectual thought.

St Paul's College, founded in 1594 and often cited as the first Western-style university in the Far East, played an important part in training seminarians for Christian missions to other parts of the region. Priests, and others in the Church, also exerted considerable political influence through the ages, in their capacity as leading figures in Macao's society and members of its educated elite.

Despite its influence, Christianity has been far less successful in Macao than one might imagine. The majority Chinese population largely adheres to an eclectic mix of religious practices, including elements of ancestor worship, Confucianism, Taoism, and Buddhism. The fishing communities, who were already based along Macao's shores when the Portuguese arrived, had their own distinct religious beliefs and practices. Indeed, Macao is thought to have taken its name from a corruption of the Chinese words 'A-Ma Gau', meaning the Bay of A-Ma, a goddess worshipped by the fishing community. An early temple to A-Ma on the tip of Macao's coastline is still an active place of pilgrimage.

Today, Catholics make up less than 10 per cent of the population. The number of worshippers has risen and fallen in tune with the various waves of emigration of the Macanese community from Macao. Some estimates suggest, for example, that about a third of practicing Catholics emigrated in the late 1960s, a time when many Macanese left, unsettled by the impact of the 1966 riots and the outpouring of anti-Portuguese feeling. Since then, the numbers seem to have recovered at least to some extent, swelled by converts from within the Chinese community— a social shift that has subtly altered the composition of local congregations.

The present figure could face further erosion with the increasing number of new arrivals from mainland settling in Macao. The Church still exerts considerable influence; it provides about half the enclave's schools (one estimate is that the Catholic church is now providing education for some 30,000 pupils), and runs many of the charitable institutions.

The post-handover constitution, the Basic Law, guarantees that religious diversity will be tolerated in Macao, but it remains unknown whether a general decline of Portugal's cultural influence will also lessen the influence of the Church.

Some moves towards localization have already taken place. The first Chinese Bishop of Macao, Domingos Lam, was appointed in 1990. In contrast to the growing number of local Chinese priests, the number of Portuguese priests working in Macao has steadily dwindled. Those who leave or die are now less likely to be replaced with new priests arriving from Portugal.

A debate is also underway about the structure of the Church hierarchy, and whether it would now be appropriate, for example, to replace the Portuguese language with Chinese language, in some areas where Portuguese language is still used in administration and worship.

While some priests seek to embrace the traditions of the Catholic church in Macao's past, others want the style of the Church to evolve in post-handover Macao, in the hope of making the institution more accessible to, and representative of, the local majority of Chinese worshippers.

Macao's population profile and local culture are undergoing rapid change, made all the more acute by the political transition now gripping the enclave after more than four centuries of Portuguese administration. As the different views expressed by the religious figures in this section illustrate, the role of the church—and Macao's long held status as a City of God—are now facing one of the greatest processes of change since the missionaries settled in Macao centuries ago.

Father Luis Sequeira

Jesuit Leadership

It is difficult to discuss Macao's historical development in the past four centuries without talking about the Jesuits. Jesuit missionaries and scholars came out to Macao in the earliest days of Portuguese settlement. As well as founding an influential intellectual centre, they also fulfilled an important role as de facto ambassadors, at a time when, in Europe, religious seniority often gave considerable political power.

The Jesuits comprised an educated elite who could represent the interests of the growing number of merchants as they tried to negotiate at a diplomatic level with Chinese officials in the sixteenth and seventeenth centuries. As a result, the Jesuit community soon became a dominant force in education, welfare, and politics in the fast-developing port.

St Francis Xavier, an early Jesuit missionary, pioneered attempts to introduce Christianity to China in the sixteenth century—without success. He died on Sancian (Shangchuan Dao), an island in the South China Sea close to China's coast, on 3 December 1552. Many other Jesuits, fathers and brothers, followed in his footsteps, and by the seventeenth century, Macao had become a centre of learning and Christianity, whose influence stretched across the whole of the Far East.

The riches earned by Portuguese traders were translated in part into splendid churches and St Paul's College, one of the greatest European institutions in the region at the time, and an important centre of learning and faith. The ruins of the Church of the Mother of God, built on the same site in 1602 and destroyed by fire in 1835, are one of the best-known symbols of Macao.

Today, the Jesuits still provide educational and social services, as well as a focus for religious practice. They remain particularly active in local culture. Father Luis Sequeira is the Superior of the Society of Jesus in Macao. Its numbers in the enclave have now declined to fewer than a dozen priests. The Jesuit Seminary in Macao now has close ties with mainland China, as well as Hong Kong and Taiwan.

Father Luis was sent to Macao by his superiors in 1976, after completing studies in humanities and philosophy in Portugal. He was only twenty-seven years old and the posting to the Far East came as a complete surprise. From 1976 to 1982, he studied Cantonese at the Chinese University of Hong Kong, and completed his basic training in theology at the Jesuit Seminary in Hong Kong. In 1982, he was ordained a priest in Macao.

I WAS SENT TO MACAO TO FULFIL A MISSION HERE. When my superior broke the news that I should go to Macao, I remember saying: 'I don't see why I have been chosen—but if I have to, I will go. There will be difficulties and pain but I will go with an open heart.' I had never in my life thought about going to the Far East.

The man who ordained me in Macao was Archbishop Dominic Tang. He had spent twenty-two years in prison in China. It was his first ordination. I felt that was an important symbol of the unity of the universal church—a Chinese bishop ordaining a foreigner. In the universal church, there are no boundaries of race or nationality. I went to Rome for a few years to complete my studies in theology and basically settled in Macao from 1985.

When I first arrived, I was constantly surprised by things I saw. Take Sundays, for example. In Europe, during my childhood at least, we celebrated Sunday quite respectfully. The shops and banks closed, and work stopped on all construction sites. Here, Sunday is a day like any other day. I found that really shocking, that absence of a religious atmosphere on Sundays.

Even on a superficial level, there were so many differences. In Europe, if you want to guide someone who is reversing a car, you shout: 'Go, go, go', and when you want them to stop, you bang the back of the car with your fist. Here, they do it the other way round. They bang repeatedly on the car to mean 'keep going', and stop banging when they want you to stop. Little things like that made me realize I was in a different world.

The manners here seemed more delicate and gentle than the Latin culture, where people are very outspoken and expressive. Once, I arrived late for the jetfoil and my companion asked the jetfoil manager to wait for me because I had forgotten my passport. And they did! They sat there and waited while I rushed all the way home and collected it. It was a gesture of respect because I was a member of a religious order, a man of the church. That really impressed me.

I don't know how much my superiors foresaw when they sent me off to Macao all those years ago, but the sense of a wider China mission has developed enormously from this small place, Macao. First, there was a union of the two Jesuit communities in Macao in 1987, then, in 1993, a union of Macao and Hong Kong as a Jesuit unit, and slowly a union with Taiwan and with our companions in China as well. I see us moving more and more towards the unity of China under one country, two systems, and I see myself as a liaison figure in that process. History has brought me a far wider and more challenging horizon than anyone expected.

Religiously, the work is a continuous journey of discovery. When I came here, I wasn't familiar with Taoism and Buddhism, but I have tried to consider them with respect—and even a certain awe and amazement. I see in Chinese culture a great sense of sharing the inner intrinsic desire of everyone to find God. I try to accept the differences between our religions, but the true revelation of God, I believe, is revealed in Jesus Christ. I enjoy making comparisons. It's a constant tension between two peoples, two religions, two cultures.

My own work here has three different dimensions. First of all, I am a priest and the head of the Jesuits in Macao. We have a variety of people here—Chinese, Italians, Spaniards, and Australians. The fundamental tradition of the Jesuits is to guide retreats—three, eight, or thirty days of spiritual experience—here in Macao or, increasingly, elsewhere in Asia. Much of my work involves that priestly ministry. In my second role, I am the principal of a school in Macao with almost two thousand students, so I have a hand in education. And thirdly, although I'm not a scholar by profession, I feel strongly about cultural issues so I have become involved in a range of historical and cultural activities in Macao that involve using the media—radio, television, and the press—where possible.

I feel very positive about Macao's transition to China. The concept of one country, two systems implies sovereignty and respect for the law, but at the same time we can add our own experience to China. Once again, Macao can be a door or a bridge to the outside world. The Chinese are very practical—they want to see before they believe. I think we have to show them that a new relationship is possible.

There is still a great block in thinking between China's church and the universal church, based on previous experiences. It's a sort of split. They mix up the idea of the Vatican as a political entity and not a spiritual reality. I think our task now is to break down that block by showing we respect the sovereignty of the Chinese government, instead of trying, as the superpowers did in the nineteenth century, to humiliate China politically—something which really damaged their understanding of the universal church.

We need to concentrate on helping China develop socially, or in the field of education, or even to live with the Chinese Christian community, to show our respect for China. If they see we belong to the universal Church, but also respect their sovereignty, and that we're also contributing to the modernization of China, sooner or later they will realize that religion is not patriotic and not hostile to their country. On the contrary, it's part of the highest level of human development.

We are gradually making progress. We don't go to the Mainland to preach or to hold public meetings, but we do share in their masses, their celebrations. We help with their schools, always in a way acceptable to the authorities. We help people in need, the elderly, lepers, all done through the official churches.

If we respect the principle of one country, two systems, I think we can achieve a lot. Any service that is brought to China in a spirit of power and authority isn't going to be accepted, but if it's brought in a spirit of true service, of respectful collaboration, real good can be achieved. That's true in whatever field you're working—whether you want to build a big airport, or container port, or whatever. If your attitude is that you are important and know best, the Chinese people won't respond. They will wait patiently until you have calmed down and are ready to respect their authority and then they'll give you permission to go ahead.

You can apply this principle to everything, the Church too. If there's a sense, right or wrong, that the authority comes from outside, that's seen as an invasion of China's authority. When we show through our actions and through our witness that we're not preoccupied with political power, China will one day understand that believers, even a universal body of believers, are not trying to conquer China. Thinking about this helps me accept the present obstacles more calmly.

Macao has changed a lot since I came here twenty years ago. The new residential areas in Taipa are very obvious and although the old-style colonial houses haven't completed disappeared, modern buildings are replacing them. I do think we should preserve some areas of the old city with their original buildings as a way of safeguarding Macao's history. The people have changed too—there's a new frenzy, a new tension. More cars, new roads, more businesses—the pace is much faster than before. People are in more of a rush, not as gentle as before, although compared with Hong Kong, I still think Macao people are more people-oriented, more open to each other.

The influence of religion was greater before, too. Nowadays, many people have flooded to Macao from the Mainland, with a Marxist or Communist background. They've even lost the old Confucian traditions. They come here without much sense of religion, but with a desire to make money. It's a refugee survival instinct. Naturally, this influx of people has had an impact on the sense of religion generally.

At the same time, the power of the casinos and tourism related to the casinos has grown, and this is an area in which moral values are downgraded. Gambling always preys on human weakness and makes money out of that weakness—these people are being psychologically manipulated. The casinos also breed prostitution, which is an exploitation of human beings. With all this change, developing a sense of religious and spiritual values is much tougher.

On the other hand, although the Church generally isn't as influential as previously, it still has a strong influence in two areas: education and social welfare. Two-thirds of the schools used to be run by Catholic institutions. Today, that number has fallen to half. Most social welfare programmes are still run by religious institutions, and that brings a great sense of the Church's presence. Religious values are entering the population through these two routes. There is a tension in Macao's name: Macao, City of the Name of God, and Macao, City of Casinos.

Sometimes I think the casinos are more evident—but, although some people come to Macao and lose themselves in the entertainment industry, there are still those who come here and rediscover their faith, and become better Christians. This is a time of crisis, and crisis can lead either to death or to resurrection. I believe we need a time of renewal in the Christian community. Even saying this, I can see already signs of a spiritual surge taking place. The Church is attracting a lot of young people from the Chinese community. The Church in China has suffered a great deal and when people suffer, they become creative. In China, the sense of vocation to the priesthood is growing and Christian communities are also growing. Mainland Chinese Christians are tough men and women, and great believers, which could be very healthy for this old, well-established Christian community in Macao that rather needs waking up and shaking up, too.

Personally, I refuse ever to enter a casino building. That's the way I express my own viewpoint. If people arrange to meet there for dinner or lunch, I won't go. No matter how important the people are who invite me, I won't go. I am highly critical of the economic system, and its reliance on the entertainment industry, because I refuse to accept so much exploitation of human weakness. If we put this much value on money, without respecting other people and while abusing human frailties, what follows is what we've got now. It is the desire for money that has led to the violence and the powerlessness to control it.

Most of the cases of violence are intimately connected with people searching for money. The root is rotten; so too are the results. Now we're seeing increasing violence developing between these gangs who are rivalling each other for control. The root of the problem, the entertainment industry itself with gambling at its centre, isn't being tackled. What I find most painful is the thought that people have lost their sense of moral values.

I can see the police need to regain control, but I also believe it has to be a common effort. Many of the professionals in the gangs come from China, and when they run into problems here, they just nip back to China. Macao can't take all the blame. Many of the gangsters come from Hong Kong, so they can't just wash their hands of the violence either. They all come here to get a share of the action. It is a threefold responsibility, and no one government can blame the others.

I admit there is a small percentage of Portuguese people here lamenting the fact we're leaving, but the reunification of Macao with the Mainland also leads us to a new stage in China's history. It is the end of an era, but instead of looking backwards and lamenting, we should see this as a time of greater unity, and that's very positive. Historically, we lived here under a gentleman's agreement. It's different from Hong Kong's history of wars and opium. China never gave up its sovereignty over Macao—they allowed two systems to coexist side by side—so, all we're really seeing is a formalization of what was happening anyway.

In a way, I will feel I'm a remnant, being Portuguese, but there's no point in longing for times that have passed. I need to adjust myself to reality. Chinese authority will be implemented in this little part of China. The mission I was sent here to fulfil has no boundaries, and in that context, I believe China should take back what belongs to it.

There is one man who seems to me to be the supreme example of what it can mean to be a Jesuit and to be Portuguese. He was a Jesuit priest [Adam Schall, 1591–1666], who lived in Beijing in the seventeenth century, and was a very close friend of the Emperor. His epitaph on the tombstone in a Beijing cemetery was written by the Emperor himself. He was an accomplished musician; he and the Emperor used to play duets together. There are accounts of him playing the clavichord until the Emperor fell asleep. He was also an excellent mathematician and President of the Observatory in Beijing. As a delegate from China he witnessed the signing of the first treaty between Russia and China, and he was also the driving force behind China's Edict of Tolerance, which allowed religious freedom in China at that time.

A Portuguese priest who was intimate friends with a Chinese emperor—both musicians, both scientists, both diplomats—and as a priest, too, he was loyal to his own convictions. Two people with different histories and cultures sharing the highest level of understanding. It is a symbol of what is possible. At the time of transition in 1999, this is the example I would like to celebrate; a symbol of the deep connection of our two countries, our two peoples, through those two extraordinary men.

Father Peter Chung

Priest in the Community

Father Peter helped to found Macao's Catholic Pastoral Youth centre, a new style of community centre established in 1976. The Catholic Church has been involved in charitable work and education in Macao for centuries, but given the changing political climate immediately after the revolution in Portugal, and the granting of more autonomy to Macao, Father Peter and his colleagues decided it was time to offer something a little different.

The centre is not only for Christians, and does not aim to convert people to Christianity. Its purpose is to offer courses and opportunities for Macao's young people that are otherwise unavailable through semi-skilled day jobs. It also aims to develop leadership skills in local youth, Macao's potential future leaders.

Father Peter received his education in the Diocesan Seminaries in Macao and Hong Kong; later in life he completed post-graduate studies in Chicago, and at the Chinese University of Hong Kong. Much of his current work is with newcomers to Macao, arriving from the Mainland. He is also a member of the Basic Law Consultative Committee on Macao. In an attempt to raise the social and political awareness of Macao's youth, he has organized a number of public debates about the transition leading up to Macao's return to China.

MY FATHER WAS BORN IN MACAO, but after the war the economic situation was bad here, so he decided to go and work in Hong Kong where there were far more opportunities. I was born there, but when I was three, we returned to Macao, and I've lived here ever since.

I attended a Catholic primary school. In my final year there, the principal, a Jesuit Priest, asked us all who wanted to study in the Seminary. In those days three of the seminarians used to come and play football and basketball with us, at weekends. I loved football so I put my hand up and said I wanted to go—because I wanted to play more football.

My family wasn't well off and at the Seminary everything was free: food and lodging, the classes, and everything. That might have been why my father agreed. There were six of us in the family—I was the oldest child—so he had a hard time paying for all of us. My mother died when I was nine. My father remarried and brought another three children into the family with his new wife. There are advantages to being poor—we didn't fight over my father's inheritance when he died. I started at the Seminary in 1962 when I was thirteen, at the Minor Seminary, or secondary school, and graduated in 1968.

Life in the Seminary in Macao was strict and traditional. We had to study in Latin, Portuguese, and Chinese. I remember my first day there well. I'd never been to the Seminary before, and it was like a prison—dark, with narrow staircases. We were only kids, wearing shorts, and we all started together at three o'clock that afternoon. A priest came down and led us up to a big room, and another Portuguese priest came in to see us. He wasn't smiling and I got the feeling he had just got up from his siesta. It was September, and it was hot. He looked serious and a bit bad tempered. He read us some rules and dismissed our parents.

We were then led to a long dormitory. There were about fifty Minor Seminarians; we all slept together in that huge dormitory. All those beds! We were told to change into a long black cassock. We were only about four feet tall, and the cassock went right down below our knees and over our wrists. We had to wear it all the time, from 5.30am to 9.30pm, all day, unless we were doing sports. And we had to wear sandals with no socks—I'd never experienced anything like that.

The Seminary had one vast study room, where everyone sat at separate desks. Whenever we left this study room to go for recess or lunch, and returned, we had to line up one by one down a long corridor. I remember there was a prefect from the Major Seminary who kept giving us instructions in Portuguese, and I couldn't understand because I'd never learnt Portuguese. That afternoon we had to pray the rosary in Portuguese, which I couldn't understand at all—I just mumbled along with everyone else.

The dining room was also huge. We ate with the Major Seminarians, so altogether there were about a hundred people. There were Chinese,

Macanese, some Timorese, and Portuguese sent from Portugal. We used European utensils—I'd only used chopsticks before. I was interested in all these different kinds of people, but we weren't allowed to talk at mealtimes. In fact, as Minor Seminarians, we were only allowed to speak to the Major Seminarians twice a year, at Christmas and Easter. Otherwise if we were caught talking, we were punished and forced to kneel in the dining room. It was very strict.

Every first Sunday of the month, in the afternoon, was the best time. That was when our families could come and visit us. That was great. My grandmother doted on me and didn't miss one month all the time I was there. She brought me Chinese soup and roast duck, which was my favourite, and chocolates. My little brother and sister liked to come, too, so they could share all the goodies!

We boys would stand in the garden and wait for the porter to call out our number—that meant our family had arrived. We were allowed to see them for one hour. We weren't allowed to take anything back into the Seminary with us, because the Portuguese students didn't have any visitors and they didn't want to make them feel worse. We had to eat what we could there and then.

Sundays and Thursdays were holidays. On Thursdays, we had a retreat once a month and the other three Thursdays, we had outings. We went for long walks all over Macao. Each year we were allowed to leave the Seminary to go home and visit our families twice—on Christmas Day and Easter Sunday—for eight hours each time.

I think it was too harsh. We were meant to leave our families and dedicate ourselves to the service of the Church, and we accepted that at the time, but it was hard. For example, there was no hot water in the Seminary. We had very short haircuts, and no mirrors. There was a sense that physical comforts weren't good for the spiritual life, just like the philosophy of the Middle Ages. There was much emphasis on obedience and self-sacrifice, and long hours of prayer. As long as we tried our best at our academic studies, that was enough—the spiritual life was more important.

During the 1 2 3 incident, even the Seminary walls were posted with sheets of Chinese characters. They were Communist slogans, anti-Portuguese and, in fact, anti-everything and everybody. I remember someone starting to take them down but we told him not to. At the time

I didn't really understand what was happening, but now I see it was a very complicated event. I think it was a response to pressure from the Portuguese, and to pressure from China as well.

One of our priests, Father Ho, the principal of one of the Catholic schools registered in Taiwan. He fled to Hong Kong the night of the riots and never came back. The students from the pro-Communist schools stuck posters up on the walls of his school, denouncing him as a running dog.

I think the 1 2 3 incident did change things in Macao. Before it happened, the inspectors in the markets were Macanese and tended to be poorly educated. They would help themselves to the best of the produce without paying for it when they came round to do their inspections. The stall holders hated them, but they couldn't do anything about it. After 1 2 3, they stopped doing that. It gave the Chinese community a sense of unity and power. Before, the community was like grains of sand, separate and unable to stand up to the Portuguese. After 1 2 3, they formed all sorts of organizations to represent their interests, and they found they could defend themselves.

The Major Seminary in Macao offered philosophy and theology, but when I was ready to go there, it had already closed following the 1 2 3 incident. The rector had to decide where to send us instead. Some went to Portugal, and some went off to Taiwan, but there was concern that if we went to Taiwan, we might not be allowed to come back again. There was a lot of talk about Macao being taken over by Communist China. Finally, he decided to send us to Hong Kong, to the Aberdeen Seminary, and I spent six years there.

Those years were a time of great change for the Catholic church worldwide. Macao had been insulated from it because of the Cultural Revolution, but in Hong Kong I really felt those changes. Seminary life was much freer than in Macao. The daily schedule, for example, was less regimented. We could even have our own timetable.

I think I found my religious faith in Hong Kong at the Major Seminary. I met priests there whom I really admired. It wasn't their intelligence, but the way they lived their lives that made me respect them. To me, it was a fresh understanding of the idea of sanctity. In the Minor Seminary we weren't allowed to make close friends. We weren't encouraged to talk about our inner self—our feelings and ideas—to other people. It was only in the Major Seminary that I made really good friends. In the Minor Seminary, I learnt rules and obedience, but in the Major Seminary I learnt about friendship and mutual respect.

I graduated in 1974 and was ordained as a priest in Macao in 1975. During that time, I found Macao still hadn't taken off economically. There were many factories but no university. The highest level of education was secondary school. Most young people ended up working in the factories, doing hard manual work. At the same time, a number of migrants began arriving in Macao, towards the end of the 1970s, as China started to open up.

Young people here had no chance to express themselves. Many were talented, potential future leaders, but the local institutions—political, economic and social—stifled their skills. They were just working in factories all day. If they happened to be the eldest in the family, they had to sacrifice any chance of education and start full-time work at fifteen or sixteen years old, to help support the family.

We started this centre in the hope of helping them. We offered programmes, courses, and workshops, and we helped them to experience something more than the inside of a factory. We also organized small group exchanges with Hong Kong and Taiwan to give them a glimpse of how other people lived. I think it has proved a success. Some of our former friends are now taking leadership positions in Macao. Ng Kuok-cheong, for example—he was in one of our groups.

We also started courses to train volunteers in Macao, which was a pioneering idea at the time, and started voluntary work schemes. For example, we helped primary school children with extra classes, or gave special support to children of recently arrived migrants from the Mainland. We also provided out-of-school activities, organizing games, drama, or sports. Later, the Communist institutions started the same programmes and asked us to help them organize courses. We helped each other. We knew our ideology was different—but the aim of helping local people was the same.

I see all this as my duty as a priest. Being a priest isn't just about serving mass and hearing confession—if you do those things without human experience, they're nothing, just rituals. Being a priest is about life experience, about helping other people to grow and realize their potential. To be with God is a blessing, but to be with other people is a blessing too.

The Church in Macao is still very colonial. It still uses Portuguese as the official language. My assignments from Bishop Lam are written in Portuguese. Even now! I'm Chinese and so is he, but his instructions come in Portuguese. The Church hasn't been localized.

At every Easter vigil in the Cathedral, the most important and obvious sign of the Catholic church in Macao, the liturgy is said in Portuguese. I asked the last bishop why. He was Portuguese, but he could say mass in Cantonese. I asked: 'Why don't you suggest celebrating Christmas in Portuguese and Easter in Chinese so Chinese people can have this experience of participation with their own Bishop?' No response. Even now, our own Chinese bishop does the same. I'm not saying I want to get rid of the Macanese and Portuguese—but where is our identity? Where is our way of expressing ourselves as a Church? In Macao, there are two distinct Catholic communities—the Chinese and the Portuguese. They're as different as day and night.

I don't know how Beijing will react to the Church here. No one knows. I think maybe in ten years' time the situation will have changed. Right now, the leaders in Beijing are still veterans from the Long March, they're the older generation. The younger generation might be just as rigid about the Church. If people in the Church, here in Macao, live up to their ideals and do good for people, this will also help us to move on.

I think if the Church tried to be more evangelistic—not in the sense of converting people to Catholicism, but serving the people and changing their lives for the better, morally, economically, and politically—I think it would help the government to understand the role we want to play and we could cooperate. In the meantime, we make sure we do things openly. The Communist regime is afraid of secrecy, but if you make sure you act publicly, there's less problem.

Many of our priests are rather conservative. I find many people in Macao don't see any path towards change in the future and want to keep everything static. It's hard to push people in the Church to move forward. The Roman Catholic Church here has a very long history. It came with the Portuguese merchants and has been considered colonial. It is seen as a Western religion by most Chinese people. Because the Church is so close to the Portuguese government, maybe the Church is a bit tense as we approach the political change.

The present bishop, Bishop Lam, is the first Chinese Bishop here in four and a half centuries. Before him, they were all Portuguese. He was only consecrated as bishop in 1988, after the decision to return Macao to China. He is under a lot of pressure from China, from Portugal, and also from Taiwan and the Vatican. It's not an easy position to be in. I think he wants to keep things stable without too many changes.

Some of the older generation don't trust the Communists. Many of them, or their parents or grandparents, have experienced Chinese political movements; not only the 1 2 3, and the Cultural Revolution, but even before that. And many of them escaped from China themselves, in the 1950s. They know what it can be like. The younger generation has no sense of that history. The schools don't teach students about it, and they have little real idea about politics on the Mainland. They feel: 'We're all Chinese—we want to share our beliefs with other people in China.' They want to be reunited with the motherland—to know the country and make a contribution to its unity.

We sent a group of young people to the Mainland last year to help in a village, offering tuition and playing with the children. In the evenings, they gathered the Catholic children together for Bible stories. There are some deeply religious villages in China, but they don't want to express their beliefs openly because of the political situation. The children have only memorized prayers and stories passed down to them by their parents. They don't have much sense of the deeper meaning of these beliefs. Our young people shared with them—and found it very meaningful. We tell our volunteers not to do too much of this evangelizing—but to experience village life in China, and teach their faith by their example of living it, rather than by preaching.

These differences between the older and younger generation, and all these political pressures, are on the first Chinese Bishop's shoulders— yet he's the only one who can lead us forward.

In Macao, a lot of people have forgotten all the old differences between the Portuguese, Macanese, and Chinese communities, and there's a sense of all of us waiting for the handover together. Life has a funny way of coming full circle. I heard that the Macanese complained recently to the Select Committee that their names had been wrongly translated from Portuguese into Chinese characters. When the Macanese fight for their identity, perhaps they'll remember what they did to the Chinese in the past. That when we went to their registry or to get our ID cards, they just handed us a Portuguese name—so that someone's father's and brother's names might come out differently from their own. Now they know what it feels like.

Photograph courtesy of the Historical Archives of Macau

Father Lancelot Rodrigues

Riding the Refugee Wave

Father Lancelot is a well-known character in Macao, famous for his charity work and down to earth approach. His fondness for cigarettes, singing, and good whisky have been well documented—along with his extrovert friendliness.

His care of the changing population of refugees has earned him much respect. His role began soon after the end of World War II. Macao's population swelled to half a million in the late 1940s, during China's civil war and after the establishment of the Communist-led government in 1949. Macao, forced to establish makeshift refugee camps, housed tens of thousands of new arrivals, many of them poor and escaping from the Mainland with few belongings. Father Lancelot, then a newly ordained priest, ran one of these camps. He lived there alongside the refugees, most of whom were then sent to third countries for resettlement.

The camps for Portuguese refugees from Shanghai closed in 1963—but the refugee problem reappeared in the late 1970s, when the first Vietnamese refugees began to arrive in Hong Kong and Macao, escaping the political turmoil in Vietnam. Their numbers soon swelled to thousands. Father Lancelot was called upon again to run the camps.

Father Lancelot came to Macao from Malacca at the request of the Bishop of Macao, with the consent of his parents, to study for the priesthood. He arrived at the age of twelve, with three other boys—though he alone went on to realize his vocation, and was finally ordained as a priest after studying for thirteen years. One of his main activities now is running charitable programmes inside China.

WE ARRIVED ON A SLOW BOAT from Hong Kong. My first memory of Macao was of seeing its huge mountains rising out of the water. In those days, Macao still had an inner harbour, and I remember the students from the Seminary coming down to meet us three new boys accompanied by a priest. I remember the jetty, all the noise, the shops, the crowds of people.

The students looked me up and down somewhat suspiciously, because I was only wearing shorts and a big sweater covering them up. I think they were wondering if, coming from Malacca, we had monkey tails, and they were straining to see if I had anything on under my sweater! We were led to the Seminary up a huge flight of steep steps. It was beautiful on the way up—trees and birds everywhere. Then we saw the building at the top—a huge, gloomy thing. I said to myself: 'Oh my goodness, is this it?' In my tiny mind, I was really scared, but they made us comfortable.

We were given little cassocks, which we had to don every day and every night. In summer, it was terribly hot. I went through all the studies, laboriously. We had to learn Portuguese, Latin, theology. We had a lot of exams. We played sports, soccer mostly, but most of our time was for studying. We were under the Jesuits and the discipline was very strict. I failed once, in Latin, in the second year, because I hadn't studied enough, but I got through all the theology and philosophy studies. I was ordained a priest in October 1949.

That was the year the first batch of Portuguese refugees came to Macao from Shanghai. The government had to look after them because they were Portuguese nationals. I was appointed by the Bishop to deal with them. So off I went, to try to cheer them up, and help them forget about the past, which wasn't an easy task. We were domiciled at the kennels built for greyhound racing. At the time, the racing was defunct so there was plenty of space for us. Even the chapel was in the kennels.

The refugees had been born in Shanghai, but had been expelled from China because they were foreign nationals. They were all mostly Macanese- and English-speaking, which made communication easier for us. There were also a lot of Chinese refugees, but the Christian charitable organization, Caritas, looked after most of them.

I was still a young man then, only about twenty-five. They were happy years, but very difficult ones too. There I was, a young fellow surrounded by so many beautiful girls! They were fine people, very well educated, and very respectful. We worked together and naturally friendships grew as we all got to know each other. We were all in the same boat.

I felt like a refugee as well. After all, I had come all the way from Malaya, and there I was, living in the camp alongside them. I had a room in the camp between the Sisters and the orphans so we could get together with the refugees every day. We couldn't abandon them. I lived there for fourteen years. The conditions were bearable. You can imagine—a converted dog house! If you convert a dog house into living quarters, there's very little privacy. They had to fill in the drains, and we made do. It was cold in winter and unbearable in summer. There was no air conditioning of course. I had to suffer with the others.

They were human beings, with a body and soul, and we had to do the best we could for them. In fact, it was an education for me, a great education, having such close contact with all sorts of people. In that camp there was a whole society, from very intelligent people, fellows who had good jobs in Shanghai, to the riffraff, all in the same basket. The foul language! And poor elderly women listening to it all every day. I learnt a lot of lessons in that camp and I learnt them the hard way.

I remember one inveterate drunken fellow in particular. He used to come to church on Sunday and I'd say: 'For goodness' sake—not you!' He'd say: 'But I have to amend'—and I'd say: 'Then bloody well amend your life!' The following week he came along again and I said: 'Tony— what happened to your conversion?'

He was the one who killed my dog. He came into the camp late with some friends, drunk, and my dog started to bark at them, knowing they were rowdy. He had a knife and killed the dog, an Alsatian. It was a real pity.

Tony passed away in the camp. He was drinking so much, even early in the morning, with an empty stomach, drinking Chinese wine. I used to say: 'No Tony, take it easy.' He was a great writer. He used to write the skits for the Saturday concerts and the songs. He was a well-educated Shanghai Portuguese, who spoke perfect English.

In the camps, you always get unrequited love and all that stuff. We had one soldier boy who fell in love with one of the women, but she didn't want him. They didn't even go out together. One evening she came out of the church and he shot her. He was sent back to prison in Portugal and the poor girl, she died. All things happen in a camp. The most terrible things, and wonderful things too.

We showed films in the camps, and every Saturday night we put on a show, and performed skits, and sang songs. They'd play their guitars, sing along. We had the grandstand, the dog racing grandstand, so the stage was right there for us. Sometimes we went for picnics to the islands and the men would swim. We had to do that, otherwise people would fossilize in the camp, doing nothing.

Many of the new arrivals were malnourished, the Chinese especially, but in good spirits. The Shanghai Portuguese refugees were well-to-do middle-class people. Macao's population went from 170,000 up to 500,000. The United States, through Catholic Relief Services, sent basic supplies, oil, and milk powder, and that sort of thing, for the schools especially, and the poor families. We had plenty of food, especially for the children. We baked bread, and made noodles every day with the flour, and gave it out to the children. They were hard times, very challenging for us all.

A refugee camp is always a terrible place. Little faults are magnified, fighting goes on. I was determined to close the camp, so I appealed to the American government to give us some money through Catholic Relief Services to build a seventy-unit block to house them and their families. We managed to get money from them, from Oxfam, and from the Macao government as well, and we built the building.

Macao was always a haven for refugees—we accepted everybody. Many went on to Hong Kong, but others went to Indonesia and Singapore. We had thousands of people passing through, using Macao as a stepping stone to somewhere else.

We finally got settled, and things calmed down, and then in came the Vietnamese refugees. When the Vietnamese came, I said: 'My gosh, not again!' The first batch of refugees came in November 1977, twenty-three of them, and twenty-four were resettled—a baby was born. They all went on to the United States, but that was just the start. The numbers grew, and after a while we were handling 15,000 to 20,000 people.

Many of them we just couldn't accept because our camps were full. Macao is a small place. Otherwise we would have had the same problem as Hong Kong, we would have been overwhelmed. But we did our best to settle people abroad. We had three camps this time. The government gave us a former barracks and we built the major camp there, and then had two other sites lent to us by the Church.

Again, camp life was terrible. We had a school full of children from the camp and the government gave us all the facilities we needed, medical support, and maternity wards. Many children were born here. The refugees were able to work, we didn't have to subsidize their daily income. They even had their own kitchens and did their own cooking. Our main task was looking after the elderly and the children, especially when they were sick.

We had a lot of fun—and a lot of fights, a lot of drunkenness. We put a stop to that. The drunkenness was because some people weren't working. At one time, the Chief Security Forces here said only those with Macao ID cards could work, even though these people had ID cards issued by the police and by the United Nations. I said to them: 'Do you fancy taking over the camp? You see the drunkenness, the fights going on here?' They relented after a couple of months. They could drink at night, if they had the money to pay for it.

The Vietnamese refugee camps I ran were entirely open. We kept law and order there. People went out, worked, came back by eleven o'clock, and if they wanted to stay outside for a night or two, they had to get permission. If they came in late without permission, they were fined. There's nothing like hitting their pockets.

I only had five who disappeared. A couple of them returned after a while and I said: 'We've been looking everywhere for you—your sister's been calling from the United States and the American government says they will give you a visa.' They said: 'All right, we'll come back,' but they never did. I had to report them to the authorities, of course. You always get one or two people who won't accept discipline. But most of them did.

There was no trouble between the refugees and the local people. The only real trouble we had were fights between the Northern Vietnamese and the Southern Vietnamese in the camp. Many of the Vietnamese married local girls from Macao, often social workers. These women went with them to another country when they were resettled. I was very glad. That way we killed two birds with one stone—gave them a job and got them married!

Compared with Hong Kong, our numbers were much smaller. It was manageable. We did have some riots, some troublemakers, but we sent them away—to Hong Kong, as a matter of fact! I closed the last camp in

1991 still knowing I had about 150 refugees in there. I took a chance, and lo and behold, two weeks later the French Government said they'd accept fifty of them, Denmark and the United States said they'd accept some, too, so I was only left with about thirty. In fact, they also disappeared.

Christianity has been here since the beginning—since Macao was first settled. Even now, you can see its importance. For starters, we have the schools in our hands, and most of the social services are run by the Church. The Sisters, for example, are committed to these things, dedicated. They're not 'nine-to-five' people.

Of course, most people in Macao aren't Catholics, but people still want their children to go to Christian schools, whether they are Christians or not. The important thing about the schooling is to give the children a good character education. In our schools, they learn a range of things, they have contact with foreign teachers. Christianity is, if anything, more important nowadays in Macao than it was when I first came here all those years ago, because now it's free. If you want to become a Catholic, a Christian, it's up to you. We don't force anyone to convert. Faith isn't a gift from your parents, it's something you have to discover for yourself.

We don't know, exactly, what the religious climate will be like after the handover, but the Chinese authorities have given us permission to carry on as we are, keeping up our contacts with outside organizations and even the Vatican. Obviously we hope that will be put into practice. The religious processions will carry on just the same too. We're a bit different from Hong Kong because we have two big processions through the streets and hopefully we will continue to hold them.

We hope they don't put limitations on anything. People must be free to believe, free not to believe. That's a basic human right. Otherwise, there are many advantages to being in Chinese territory. It's a great culture. The Chinese festivals, Chinese New Year, the Lantern Festival, and so on, mean something to us all. They must remain, but hand in hand with our culture.

I don't feel anything about the handover. For me, it's going to be another day—so what? We've already started working over the border, since 1985. We started a six-day course for the mentally handicapped and those with speech handicaps. Our experts are priests and nuns who studied in England and Australia. The authorities in the province liked it so much, they sent word to other provinces, and they have started asking us too. We get invited here, there, and everywhere.

When we saw news of earthquakes I appealed to our international donors for help, and we went to build schools, clinics, and houses in Yunan Province. We donated and distributed a lot of clothing there. We saw a lot of delapidated schools, the consequences of the Cultural Revolution, and we're now building about ten new schools in the mountain areas. We're also trying to help with water, providing hand-pumped wells. We don't preach—we go to do social work, and they know that. We visit the bishops, and sometimes they're allowed to invite me to say Mass in their church, but we say our prayers in our hotel rooms. We don't want to get into trouble with the authorities there. China is changing. When we go, the whole village comes out—they celebrate and give us a real red-carpet welcome. All that festivity just for one European devil coming to see them!